PROBLEMS IN THERMAL CHEMISTRY

PROBLEMS IN THERMAL CHEMISTRY

By
Dr. Syed Aftab Iqbal
M.Sc. , Ph.D., FICS
FICC, FIAEM, MNASc.
Professor
Department of Chemistry
Saifia Science College
Barkatullah University
Bhopal (India)

DISCOVERY PUBLISHING HOUSE PVT. LTD.
NEW DELHI-110 002

Published by:
Tilak Wasan
DISCOVERY PUBLISHING HOUSE PVT. LTD.
4831/24, Ansari Road, Prahlad Street
Darya Ganj, New Delhi-110002 (India)
Phone: +91-11-23279245, 43764432
Fax: +91-11-23253475
E-mail: parul.wasan@gmail.com
info@discoverypublishinggroup.com
discoverypublishinghouse@gmail.com
web: www.discoverypublishinggroup.com

First Edition: **2011**
ISBN: 978-81-8356-816-6

Problems in Thermal Chemistry

Printed at:
Mehra Offset Press
Delhi

Preface

The book "Problems in Thermal Chemistry" has been written to meet the requirements of graduate students of all Indian Universities. We have been selected sufficient theory covered the syllabus from Universities examination papers.

We have tried to our best to keep the book free from the misprint. The author shall be grateful to the readers who point out errors and omissions which inspite of all care might have been there. I have spared no pains in applying my long experience of degree class teaching to make the book useful to the students.

We shall indeed be very thankful to our colleague for their recommendations this book of for their students. The present book will be warmly received by the students and teachers.

—Author

Preface

The book "Problems in Physical Chemistry" has been written to meet the requirements of graduate students of all Indian Universities. We have [illegible] sufficient theory covered the syllabus of all Indian Universities [illegible]

We [illegible] [illegible] [illegible] the students.

We [illegible] [illegible] book will be warmly received by the students and teachers.

CONTENTS

1

BASIC PROBLEM OF THE THERMAL CHEMISTRY

INTRODUCTION

The heat of reaction is defined as the amount of heat absorbed or evolved at a given temperature when the reactants have combined to form the products as represented by the balanced chemical equation. If the heat of reaction is denoted by q, then according to the first law,

$$q = \Delta E - w$$

where –w is the work done by reacting mixture when it reacts and D E is the energy difference between the reaction products and the reactants,

$$\Delta E = \Sigma E \text{ (products) } \Sigma E \text{ (reactants)}$$

The numerical value of q depends on the manner in which the reaction is performed. For the two methods of conducting chemical reactions in calorimeters, we have

Constant volume : w = 0, and qv = DE; Bomb calorimeter.

Constant pressure : $w = -P\,\Delta V$ and $qP = \Delta E + P\,\Delta V = \Delta H$; Open calorimeter.

Thus heat of a reaction at constant volume (qv) is the change of energy (ΔE) accompanying the chemical reaction. The heat of reaction when measured at constant pressure (qp) is equal to the enthalpy change in the reaction given by

$$q_p = \Delta H = \Sigma H \text{ (products)} - \Sigma H \text{ (reactants)}$$

Constant pressure are much common in chemistry therefore usually whenever we speak of heat of reaction in implies enthalpy change.

THERMAL CHEMISTRY LAW

Lavoisier and Laplace's Law

This law was given by Lavoisier and Laplace in 1780 and may be stated as follows : "The amount of heat supplied to decompose a compound into its elements is equal to the heat of formation of that compound form its elements"

In simple words, the heat of formation of compound is numerically equal to its heat of decomposition of a compound but of opposite sign. Thus, the larger the heat of formation of a compound, the greater will be its stability.

Importance : The importance of this law is that thermochemical equations can be reversed. For example, the equation

$$C(s) + O_2(g) \rightarrow CO_2(g)$$

$$\Delta H^o = -\ 296.64 \text{ kJ}$$

may be written as

$$CO_2(g) \rightarrow C(s) + O_2(g)$$

$$\Delta H = +\ 296.64 \text{ kJ}$$

This law applies not only to reactions involving a compound and its component elements but also to reactions of all other types. Thus, for example,

$$C_2H_5OH + 3O_2 \rightarrow 2CO_2 + 3H_2O$$

$$\Delta H^o = -\ 1366.95 \text{ kJ}$$

Hence by Lavoisier-Laplace's law, we get

$$2CO_2 + 3H_2O \rightarrow C_2H_5OH + 3O_2$$

$$\Delta H^o = +\ 1366.95 \text{ kJ.}$$

HESS'S LAW OF CONSTANT HEAT SUMMATIONS

Most calculations involving heat changes in thermochemical equations are based on the fundamental law enunciated by Russian chemist G. H. Hess in 1840. This law can be stated in words as follows :

"The enthalpy change of a given chemical reaction is the same whether the process takes place in one or several steps".

In other words, the law is merely a corollary of the law of conservation of energy which implies that heat of reaction does not depend upon the nature of intermediate products but depends only on the initial reactants and final products and is also independent of the path or the manner by which the change is carried out.

Experimental Illustration : In order to bring about experimental illustration, we can carry out the same chemical change in two alternative ways and show that the enthalpy change remains same for both. For example, carbon dioxide may be produced by two process :

First Procedure : The burning of carbon can take place in a single step to form CO_2 with the enthalpy change which is equal to – 393.0 kJ mol^{-1}.

$$C(graphite) + O_2(g) \rightarrow CO_2(g); \Delta H = -393 \text{ kJ} \quad ...(1)$$

Second Procedure. It is also possible to carry out the above reactions in two steps. First of all carbon can be converted into carbon monoxide. Then carbon monoxide can be oxidized to carbon dioxide. Thus,

$$C(graphite) + 1/2O_2(g) \rightarrow CO(g); \Delta H = -110.5 \text{ kJ} \quad ...(2)$$

$$CO(g) + 1/2O_2(g) \rightarrow CO_2(g); \Delta H = -283.2 \text{ kJ} \quad ...(3)$$

On adding equations (2) and (3), we get

$$C(graphite) + O_2(g) \rightarrow CO_2(g); \Delta H = -393.7 \text{ kJ} \quad ...(4)$$

From equations (1) and (4), it follows that the enthalpy change in both the procedures is almost same. The small difference can be attributed to experimental error. It also follows from the above discussion that thermochemical equations can be added or subtracted as ordinary algebraic equations.

Theoretical Justification : (Hess's law as a corollary of First Law of Thermodynamics). The law also follows as a mere consequence of the First Law of Thermodynamics. Suppose a substance A can be changed to Z directly.

$$A \rightarrow Z + Q_1 \quad ...(5)$$

where Q_1 represents the heat evolved in the direct change. If the same change is carried out through intermediate stages,

$$A \rightarrow B + q_1, B \rightarrow C + q_2, C \rightarrow Z + q_3$$

The total evolution of heat = $q_1 + q_2 + q_3 = Q_2$

By Hess's Law, $Q_1 = Q_2$.

If it is not so, suppose $Q_2 > Q_1$. Then by transforming A to Z through stages and retransforming directly back to A there would occur a gain of heat = $Q_2 - Q_1$. If we repeat this process again and again, an unlimited heat will be produced in an isolated system. This goes against the First Law of Thermodynamics. Hence Q_1 must be equal to Q_2 (Fig. 1)

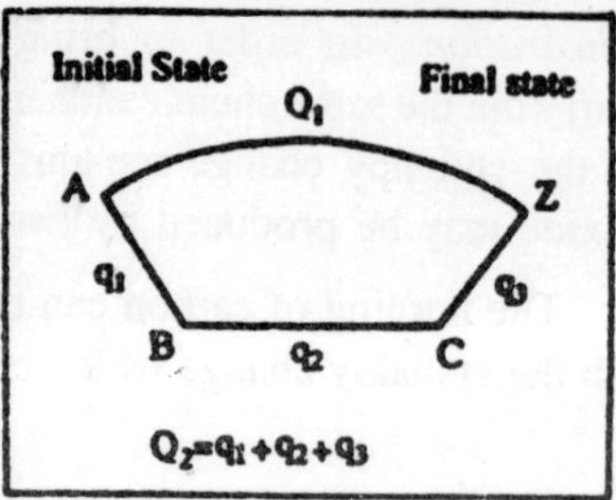

Fig. 1 : Theoretical Justification of Hess's law

APPLICATIONS OF HESS'S LAW

The main practical use of Hess's law lies in the fact that it makes legitimate to add up or subtract thermochemical equations as algebraic equations. This gives rise to the mathematical method of calculating :

(1) Heat of Reaction,

(2) Heat of Transition,

(3) Heat of Formation, and

(4) Lattice Energy of a Crystal.

Let us describe these one by one.

The importance of Eq. (5) is that the amount of heat absorbed at constant volume can be identified with the change in the thermodynamic quantity.

(1) *Calculation of heats of reactions :* By Hess's law it is possible to calculate heats of many reactions which cannot be determined experimentally. For example it is very difficult to measure the heat evolved when carbon burns in oxygen to form carbon monoxide,

$$C(s) + 1/2O_2(g) \rightarrow CO(g); \Delta H = ?$$

From the Hess's law, it is evident that the heat produced in the combustion of carbon to give carbon dioxide is the same, viz., – 393.5 kJ, whether it takes place in a single step as

$$C(s) + O_2(g) \rightarrow CO(g); \Delta H = -393.5 \text{ kJ}$$

or in two steps as

$$C(s) + 1/2O_2(g) \rightarrow CO(g);\ \Delta H = x \text{ kJ (say)}$$

$$CO(g) + 1/2O_2(g) \rightarrow CO_2(g);\ \Delta H = y \text{ kJ (say)}$$

Evidently

$$x + y = -393.5 \text{ kJ}$$

The x, the heat produced in the conversion of carbon to carbon monoxide cannot be determined but y, the heat change involved in the combination of carbon monoxide and oxygen to give carbon dioxide can be measured and has been found to be – 282.0 kJ

$$x = -393.5 - y$$

$$\text{or } y = -393.5 - (-282.0) = -111.5 \text{ kJ.}$$

Hence the heat evolved during the combination of carbon to give carbon monoxide is – 111.5 kJ.

(2) *Determination of the Heat of Transition* : As transition of an element from one allotropic form to another is an extremely slow process it requires heating of one allotropic form of the element for a long period at constant temperature. Therefore, it is not possible to determine heats of transition of elements experimentally. They are, however, determined indirectly by using Hess's law. Following example will make this clear.

Suppose we intend to determine the heat of transition of sulphur (rhombic) into sulphur (monoclinic), we start with the two allotropic forms separately and measure their heats of combustion.

$$S_{(R)} + O_2(g) \rightarrow SO_2\ ;\ \Delta H = AJ \qquad ...(i)$$

$$S_{(M)} + O_2(g) \rightarrow SO_2\ ;\ \Delta H = BJ \qquad ...(ii)$$

On subtracting (ii) from (i), we get

$$S_{(R)} - S_{(M)} \rightarrow 0\ ;\ \Delta H = (A - B) \text{ J}$$

$$S_{(R)} - S_{(M)}\ ;\ \Delta H = (A - B) \text{ J} \qquad ...(iii)$$

Equation (iii) expresses the thermochemical equation for the transition of $S_{(R)}$ to $S_{(M)}$ Hence heat of transition $S_{(R)} \rightarrow S_{(M)}$ is (A – B) J.

In this particular case, $\Delta H = 2.5$ kJ.

(3) *Calculation of Heats of Formation* : The heat of formation of those compounds which cannot be determined experimentally

can be calculated by using Hess's law. This can be understood by considering the following example:

The enthalpy of formation of methane from graphite and hydrogen cannot be found directly as no chemical reaction occurs between hydrogen and graphite.

(i) C(graphite) + $2H_2(g) \rightarrow CH_4(g)$

However, ΔH for the above reaction may be calculated by the help of the following reactions :

(ii) C(graphite) + $O_2(g) \rightarrow CO_2(g)$; $\Delta H_a = -\ 393.5$ kJ

(iii) $2H_2(g) + O_2(g) \rightarrow 2H_2O(l)$; $\Delta H_b = 2(-285.9\text{ kJ}) = -\ 571.8$ kJ

(iv) $CH_4(g) + 2O_2(g) \rightarrow CO_2(g) + 2H_2O(l)$; $\Delta H_c = -\ 890.3$ kJ

From the above reactions, the enthalpy change for reaction (i) may be obtained by adding equations (ii) and (iii)

(ii) C(graphite) + $O_2(g) \rightarrow 3CO_2(g)$; $\Delta H_a = -\ 393.5$ kJ

(iii) $2H_2(g) + O_2(g) \rightarrow 2H_2O(l)$; $\Delta H_b = -\ 571.8$ kJ

C(graphite) + $2O_2(g) + 2H_2(g) \rightarrow CO_2(g) + 2H_2O(l)$;

$\Delta H_a + \Delta H_a = -\ 965.3$ kJ

On subtracting equation (iv) from the above equation, we obtain

C(graphite) + $2O_2(g) \rightarrow CH_4(g)$;

$\Delta H = -\ 965.3\text{kJ} + 890.3\text{kJ} = -\ 75.0$ kJ

(4) *Lattice Energy of a crystal (Born-Haber Cycle) :* Hess's law can be used to determine lattice energy which may be defined as the amount of heat released when the requisite amounts of ions in the gaseous state combine to yield 1 mole of crystal lattice. An example is the determination of $DH_{Lattice}$ of sodium chloride crystal which corresponds to the reaction

$$Na^+(g) + CI^-(g) \rightarrow NaCI\ (s).$$

It is possible to determine the numerical value of $\Delta H_{Lattice}$ for the above reaction directly by using Born-Haber's cycle. The following is the sequence of steps for the formation of NaCI crystals from Na(s) and $CI_2(g)$

(i) Vaporisation of Na(s) Na(s) $\rightarrow$ Na(g) ΔH_1

(ii) Ionisation of Na(g)	$Na(g) \rightarrow Na^+(g)+e$	ΔH_2
(iii) Dissociation of chlorine	$1/2\ CI_2(g) \rightarrow CI(g)$	ΔH_3
(iv) Formation of CI^-	$CI(g) + e \rightarrow CI^-(g)$	ΔH_4
(v) Condensation of	$Na^+(g)$ and $CI^-(g)$ + CI–(g) → NaCI(s)	Na+(g) ΔH_5
Net change	$Na(s)+1/2\ CI_2(g)\ NaCI(s)$	ΔH_6

On applying the Hess's law to the above sequence of steps, we get

$$\Delta H_6 = \Delta H_1 + \Delta H_2 + \Delta H_3 + \Delta H_4 + \Delta H_5$$

Except ΔH_5, it is possible to determine experimentally all of the changes of enthalpy, *i.e.*, ΔH_1, ΔH_2, ΔH_3, ΔH_4, and ΔH_6. Hence ΔH_5 can be determined from the above relation.

EXOTHERMIC REACTION

$\Delta H < 0$, ΣH (products) $< \Sigma H$ (reactants). Heat is liberated in the reaction and is transferred to the surroundings. Therefore the enthalpy of the products is less than enthalpy of the reactants from which the products are generated. Diagrammatically these may be represented as follows:

	Reactants	Products
	|	↑
	$\Delta H<0$	$\Delta H>0$
	↓	|
	Products	Reactants
Enthalpy H,	Exothermic	Endothermic

If the reaction is performed at constant temperature, it is indicated by writing TK in parenthesis to the right of Δ H (or Δ E).

The state of a product or reactant is indicated by symbols in parentheses after the chemical symbol e. g., (s) or (c) for solid or crystal, (!) for a liquid, (g) for a gas, (aq) for solution in water.

SIGN CONVENTIONS

If heat is absorbed by the system (q > 0) then the reaction is said to be endothermic and ΔE or ΔH value is given a positive sign.

If heat is evolved (q < 0) the reaction is said to be **exothermic**, and ΔE or ΔH value is given negative sign.

ENDOTHERMIC REACTION

$\Delta H > 0$, ΣH (Products) > ΣH (reactants). Heat absorbed during the reaction increases the energy contents of the products and hence the enthalpy of the products becomes greater than that of the reactants from which the products are formed.

THERMOCHEMICAL EQUATIONS

For completeness, the chemical equations along with the enthalpy change or energy change accompanying the chemical reaction are written in two ways. In the first method, the balanced chemical equation representing the actual reaction is written, and the heat change is indicated by mentioning ΔH or ΔE with proper sign to the right side of the equation, *e.g.*, when 1 mole methane burns in the presence of oxygen, 1mole CO_2 gas and moles H_3O are formed, and 890.3 kj heat is produced measured at constant pressure and 298K. This change is represented by the *Thermochemical Equation* as

$$1CH_4\ (g) + 2O_2\ (g) \rightarrow CO_2\ (g) + 2H_2O$$

$$(1)\ \Delta H\ (298K) = -890.3kJ$$

The same equation may be written in thermally balanced manner. Since the reaction is exothermic, 890.3 kJ heat is given to the surroundings, the enthalpies of products are less than those of the reactants by 890.3 kJ. If this amount is added to the product side, the thermally balanced equation is:

$$CH_4\ (g) + 2O_2\ (g) \rightarrow CO_2\ (g) + 2H_2O\ (1) + 890.3\ kJ$$

At present the former method is preferred to the latter. Endothermic reactions may be represented in a similar manner with a changed sign of ΔH or ΔE.

$$N_2\ (g) + O_2\ (g) \rightarrow 2NO\ (g) \quad \Delta H\ (298K) = 180\ kJ$$

$$\text{or } N_2\ (g) + O_2\ (g) \rightarrow 2NO\ (g) - 180\ kJ$$

KIRCHHOFF'S EQUATION

In 1858 Kirchhoff deduced certain mathematical expressions for defining the variation of heat of reaction with temperature. All these deductions are derived from the first law of thermodynamics . Consider a process,

$$A \rightarrow B$$

Let E_A and H_A be the internal energy and heat content of the system in the initial state A. Similarly, let E_B and H_B be the internal energy and the heat content of the system in the final state B. By the first law of thermodynamics, the change in enthalpy for the given process is given by the following equation

$$\Delta H = H_B - H_A \qquad ...(1)$$

But heat content is a function of temperature and pressure. Therefore, the equation (1) can be differentiated with respect to temperature at constant pressure.

$$\left[\frac{\partial(\Delta H)}{\partial T}\right]_P = \left[\frac{\partial(\Delta_B)}{\partial T}\right]_P - \left[\frac{\partial(H_A)}{\partial T}\right]_P = (C_p)B - (C_P)A$$

$$\left[\because \left(\frac{\partial H}{\partial T}\right)_P \; C_P\right]$$

or $$\left[\frac{\partial(\Delta H)}{\partial T}\right]_P = \Delta C_P \qquad ...(2)$$

where ΔC_p represents the difference in the heat capacities of initial and final state at constant pressure. In order to integrate this equation (2), we have to consider that

(i) ΔC_p is independent of temperature

or (ii) ΔC_p is dependent upon temperature.

Let us take into consideration these cases one by one.

Case I.

When ΔC_p is constant between two temperatures T1 and T_2 : If we consider C_p to be constant and integrate equation (2) between the limits H_1 at T_1 and H_2 and T_2, we get the expression

$$\int_{H_1}^{H_2} d(\Delta H) = \int_{T_1}^{T_2} \Delta C_p \, dT, \text{ or } [\Delta H]_{H_1}^{H_2} = [\Delta C_P]_{T_1}^{T_2}$$

or $$(\Delta H_2 - \Delta H1) = \Delta C_P (T_2 - T_1) \qquad ...(3)$$

This relationship is termed as Kirchhoff's equation.

Case II.

When ΔC_P varies with temperature.

It implies that equation (2) can be integrated at constant pressure

$$\int d(\Delta H) = \int \Delta C_P . \, dT$$

or $$\Delta H = \int \Delta C_P . \, dT + K \qquad ...(4)$$

where K is a constant of integration. When T = 0. then $\Delta H_o = K$. Hence equation (4) becomes as follows :

$$\Delta H = \Delta H_o + \int \Delta C_P . \, dT \qquad ...(4A)$$

where ΔH_o represents the difference in heat contents of products and reactants at the absolute zero.

But $(C_P)_A = \propto_A + \beta_A T + \gamma_A T^2$ ......

$$(C_P)_B = \propto_B + \beta_B T + \gamma_B T^2 \text{ ......}$$

$$\therefore \quad C_P = (C_P)_B - (C_P)_A = (\propto_B + \beta_B T + \gamma_B T^2 ...) - (\propto_A + \beta_A T + \gamma_A T^2 + ...$$

$$= (\propto_B - \propto_A) + T\,(\beta_B - \beta_A) + T^2\,(\gamma B - \gamma A) + ...$$

i.e., $$\Delta C_P = \Delta\propto + T\Delta\beta + T^2\,\Delta\gamma + \qquad ...(5)$$

If we insert this value of ΔC_P in equation (4A) and carry out the integration, we get

$$\Delta H = \Delta H_o + T\,\Delta\propto + \frac{T^2}{2}\,\Delta\beta + \frac{T^3}{3}\,\Delta\gamma + ... \qquad ...(6)$$

The values of $\Delta\propto$, $\Delta\beta$, $\Delta\gamma$, ... can be calculated from heat capacity date. Equation (6) is another form of Kirchhoff's equation. This equation can be used to evaluate ΔH_o if the value of ΔH and constant $\Delta\propto$, $\Delta\beta$, $\Delta\gamma$, ... are known.

Consider the following reaction,

$$n_1 A + n_2 B \rightleftharpoons n_3 C + n_4 D$$

where n_1 and n_2 be the number of molecules of reactants A and B and n_3 and n_4 be the number of molecules of products C and D,

Now $$(C_P)_A = n_1\,(\propto_1 + \beta_1 T + \gamma_1 T^2 + \delta_1\,T^3 + ...\,...)$$

$$(C_P)_B = n_2\,(\propto_2 + \beta_2 T + \gamma_2 T^2 + \delta_2\,T^3 + ...\,...)$$

$$(C_P)_C = n_3\,(\propto_3 + \beta_3 T + \gamma_3 T^2 + \delta_3\,T^3 + ...\,...)$$

$$(C_P)_D = n_4\,(\propto_4 + \beta_4 T + \gamma_4 T^2 + \delta_4\,T^3 + ...\,...)$$

The change in heat capacity (ΔC_P) is given as follows :

$$\Delta C_P = [\{(\propto_3 + \propto_4) - (\propto_1 + \propto_2)\}] + \{(\beta_3 + \beta_4) - (\beta_1 + \beta_2)\}T$$
$$+[\{\gamma_3 + \gamma_4) - (\gamma_1 + \gamma_2)\}\ T^2 + ...]\ (n_3 + n_4 - n_1 - n_2)$$

or $\Delta C_P = (\propto + \beta T + \gamma T^2 + ...)\ (n_3 + n_4 - n_1 - n_3)$

On comparing it with Kirchhoff's equation, we obtain

$$\left[\partial \frac{\partial H}{\partial V}\right]_P = \Delta C_P = \propto + \beta T + \gamma T^2 + ... \qquad ...(6A)$$

On integrating this equation, we obtain

$$\Delta H = \Delta H_o + \propto T + \beta \frac{T^2}{2} + \gamma \frac{T^3}{3} + ... \qquad ...(6B)$$

where ΔH_o is the integration constant and

$\propto = \propto_3 + \propto_4 - \propto_1 - \propto_2,\ \beta = \beta_3 + \beta_4 - \beta_1 - \beta_2$

$\gamma = \gamma_3 + \gamma_4 - \gamma_1 - \gamma_2,\ \delta = \delta_3 + \delta_4 - \delta_1 - \delta_2,$

Equation (6B) is the general expression which defines the variation of ΔH with temperature.

FLAME TEMPERATURE

When a fuel burns, definite quantities of products are formed and a definite amount of heat is liberated. This quantity of heat may raise the temperature of the products of combustion sufficiently so as to produce a flame. This temperature is termed as the maximum flame temperature, or calorific intensity.

When the experiment is carried out at constant volume the quantity of heat liberated will be ΔE and the maximum temperature is termed explosion temperature. In case the experiment is carried out at constant pressure then heat liberated is ΔH and the maximum temperature to which the products will be raised is termed as the flame temperature. Generally this temperature is termed as the adiabatic maximum flame temperature as it is assumed that there occurs no loss of heat in the process.

Calculation of Maximum Temperature

(1) If we assume that the reaction is very fast and the total heat evolved is used to raise the temperature of the system, then the heat capacities of the species present at the completion of reaction and the heat evolved can be used for calculating the maximum temperature by using Kirchhoff's equation in the following form,

$$\Delta H_{(reaction)} = \int_{T_2}^{T_1} \sum nC_P \, dT \, \Delta H_{(heating)}$$

where it is assumed that $\Delta H_{(heating)} = -\Delta H_{(reaction)}$, as the heat evolved is taken by the products. T_1 represents the initial temperature and T_2 the maximum flame temperature, $\Sigma(nC_P)$ the sum of the heat capacities of all the species present in the system after the reaction gets completed. For constant volume process,

$$\Delta E = \int_{T_2}^{T_1} \sum nC_V \, dT$$

The direct way to know the maximum flame temperature is to carry out the experiment in the laboratory under adiabatic conditions and measure the temperature using standard techniques of temperature measurements.

The calcualted flame temperature has been found to be higher than the observed temperature. This difference has been attributed to a number of complications which are as follows :

(i) The system may not be truly adiabatic.

(ii) In combustion, we generally use excess of air or oxygen, Unreacted air or oxygen is to be heated and therefore a part of the heat would be utilised in heating these gases.

(iii) When flame temperature is to be calculated, heating of the unreacted gases should also be considered.

(iv) At high temperatures, the products might decompose and a part of the heat may be utilised to decompose the product molecules.

(2) If an exothermic reaction is assumed to carry out under adiabatic conditions, the heat evolved would be able to raise the temperature of the system.

It is possible to determine the rise in temperature by considering the reaction to take place in the following two steps, both at constant pressure.

Step I : Reactants (T_o, P) → Products (T_o, P) ΔHT_o

Step II : Products (T_o, P) → Products (T_f, P)

The heat involved in the second step may be given as follows :

$$\Delta H_2 = \int_{T_o}^{T_f} C_P \text{ (Products) } dT$$

As the overall reaction refers to the sum of the above two steps, *i.e.*,

$$\text{Reactants } (T_o, P) \rightarrow \text{Products } (T_f, P)$$

the net heat change is given as follows :

$$\Delta H = \Delta H_{To} + \Delta H_2 \qquad \text{...(i)}$$

As the reaction is carried out under adiabatic conditions, it is evident that $\Delta H = 0$. Hence Eq. (i) may be put as follows :

$$\Delta H_{To} + \Delta H_2 = 0.$$

or

$$\Delta H_2 = -\Delta H_{To} \text{ or}$$

$$\int C_P \text{ (Products) } dT = -\Delta H_{To}$$

If C_ps are regarded to be independent of temperature, then

$$C_p \text{ (Products) } (T_f - T_o) = -\Delta H_{To} \text{ or } T_f = \frac{-\Delta H_{TO}}{C_p \text{ (products)}} + T_o.$$

However, the approximation that C_ps are independent of temperature has not been correct because T_f is a very large quantity ($\Delta H_{To} \sim$ kJ and CP $\sim$ J) of the order of thousands of kelvin.

If the reaction is carried out in a closed vessel *i.e.*, under the condition of constant volume, then ΔE and C_V replace ΔH and C_p respectively, *i.e.*,

$$T_f = \frac{(-\Delta E_{To})}{C_v \text{ (products)}} + T_o$$

EXOTHERMIC AND ENDOTHERMIC REACTIONS

(a) *Exothermic Reactions* : A reaction which proceeds with the evolution of heat is referred to as an exothermic reaction. Some examples of exothermic reactions are as follows :

$$2Zn(s) + O_2(g) \rightarrow 2ZnO(s) + 693.8 \text{ kJ} \qquad \text{...(1)}$$

$$C(s) + O_2(g) \rightarrow CO_2(g) + 393.5 \text{ kJ} \qquad \text{...(2)}$$

$$C_2H_2(g) + 2H_2(g) \rightarrow C_2H_6(g) + 314.0 \text{ kJ} \qquad \text{...(3)}$$

In all the above reactions, heat is evolved and therefore these are examples of exothermic reactions. If an exothermic reaction is taking place at constant temperature and constant volume, the energy change is expressed in terms of internal energy changes, *i.e.*, ΔE. Τηυσ,

ΔE = Internal energy of product - Internal energy of reactants

$= E_P - E_R$

In exothermic reactions, the total internal energy of the reactants is more than that of products and therefore the excess of internal energy, ΔE, will l appear as heat of reaction or in other words ΔE is negative for exothermic reactions. Thus, equations (1), (2) and (3) in terms of ΔE can be written as :

$$2Zn(s) + O_2(g) \rightarrow 2ZnO(S) \; ; \; \Delta E = -\,693.8 \text{ kJ}$$

$$C(s) + O_2(g) \rightarrow CO_2(g) \; ; \; \Delta E = -\,393.5 \text{ kJ}$$

$$C_2H_2(g) + 2H_2(g) \rightarrow C_2H_6(g) \; ; \; \Delta E = -\,314.0 \text{ kJ.}$$

If an exothermic reaction is taking place at constant temperature and constant pressure, the energy change is expressed in terms of enthalpy change, ΔH, *i.e.*,

ΔH = Enthalpy of products– Enthalpy of reactants $= H_P - H_R$

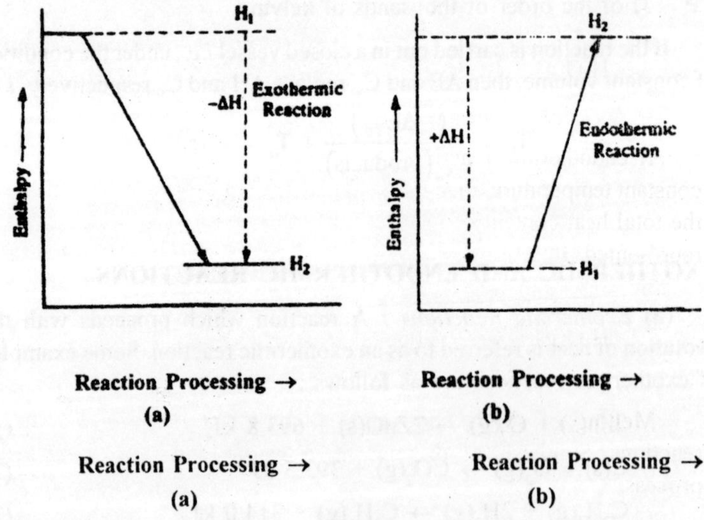

Reaction Processing →

(a)

Reaction Processing →

(b)

Fig. 2

In exothermic reactions, the total heat content of the reactants is more than that of products and, therefore, the excess of heat constant will appear as heat of reaction. In other words ΔH like ΔE is negative. Thus, equations (1), (2) and (3) in terms of ΔH can be written as :

$$2Zn(s) + O_2(g) \rightarrow 2ZnO(S) \; ; \Delta H = -\ 693.8 \text{ kJ}$$

$$C(s) + O_2(g) \rightarrow CO_2(g) \; ; \Delta H = -\ 393.5 \text{ kJ}$$

$$C_2H_2(g) + 2H_2(g) \rightarrow C_2H_6(g) \; ; \Delta H = -\ 314.0 \text{ kJ}$$

As ΔH is negative for exothermic reactions, this is illustrated in Fig. 2(a) and (b).

(b) *Endothermic Reactions* : A reaction which proceeds with the absorption of heat is referred to as an endothermic reaction. Some examples of endothermic reaction are as follows :

$$2HgO(s) + 180.4 \text{ kJ} \rightarrow 2Hg(l) + O_2(g) \qquad ...(1)$$

$$H_2O(g) + C(s) + 131.2 \text{ kJ} \rightarrow CO(g) + H_2(g) \qquad ...(2)$$

$$N_2(g) + O_2(g) + 180.5 \text{ kJ} \rightarrow 2NO(g) \qquad ...(3)$$

If endothermic reactions are taking place at constant temperature and volume, the total internal energy of the reactants (E_R) is less than the total internal energy of the products (E_P), *i.e.*, ΔE is positive. Thus, equations (1), (2) and (3) can be put as :

$$2HgO(s) \rightarrow Hg(l) + O_2(g); \text{ DH} = 180.4 \text{ kJ}$$

$$H_2O(g) + C(s) \rightarrow CO(g) + H_2(g); \text{ DH} = 131.2 \text{ kJ}$$

$$N_2(g) + O_2(g) \rightarrow 2NO(g); \text{ DE} = 180.5 \text{ kJ.}$$

If endothermic reactions are carried out at constant pressure and constant temperature, then the total heat content of products is more than the total heat content of reactants. In other words, DH is positive. It is represented. In terms of DH, equations (1) (2) and (3) may be put as:

$$2HgO(g) \rightarrow 2Hg(l) + O_2(g); \text{ DE} = 180.4 \text{ kJ}$$

$$H_2O(g) + C(s) \rightarrow CO(g) + H_2(g); \text{ DE} = 131.2 \text{ kJ}$$

$$N_2(g) + O_2(g) \rightarrow 2NO(g); \text{ DH} = 180.5 \text{ kJ}$$

Melting of solids and vaporization of liquids are endothermic reactions. Synthesis of proteins in living systems is an endothermic process.

(c) Differences between Exothermic and Endothermic Reactions.

The differences are given in tabular from :

Table 1

Exothermic Reactions	*Endothermic Reactions*
1. Heat is given to the surroundings.	1. Heat is absorbed from the surroundings.
2. The internal energy of reactants is more than that of products, *i.e.*, ΔE is – ve.	2. The internal energy of products is more that of reactants, *i.e.*, ΔE is + ve.
3. The enthalpy of reactants is more than that of Products, *i.e.*, ΔH is –ve.	3. The enthalpy of reactants is less than that of Products, i.e, ΔH is +ve.

DIFFERENT TYPES OF ENTHALPY CHANGES

Enthalpy of a reaction is given different names depending upon the types of reactions. Enthalpy of a reaction depends on temperature, a standard state is chosen to compare the enthalpies or reactions of different substances. A convenient standard state for a substance is the most stable state of aggregation under 1 atmosphere and at the specified temperature, generally at 25°C or 298 K.

For pure solids, liquids and ideal gases, the standard state corresponds to the state of the substances at 1 atm. Pressure and specified temperature.

For the dissolved substances, the standard state of the solute is that concentration which gives unit activity.

The standard enthalpy of a reaction at a temperature T and at a pressure of 1 atm is denoted by $\Delta H^o(T)$ and is the value of $\Sigma \Delta H_{products} - \Sigma \Delta H_{reactants}$, at that temperature when all the substances (or elements) in the reaction are in their standard states.

The enthalpy of every element in its standard state is arbitrarily given a zero value. For example, the enthalpy of formation of a mole of $CO_2(g)$ from its elements in the standard state is – 393.5 kg mol^{-1}, *i.e.*,

$$C(graphite) + O_2(g) \rightarrow CO_2(g);\ \Delta H^o = -393.5 \text{ kJ}.$$

The value of ΔH^o_{298K} is given as follows:

$$\Delta H^o_{298K} = H^oCO_2(g) - [H^oO_2(g) + H^o_{Cgrphite}]$$

In the above equation, H° refers to the standard enthalpy per mol.

As the enthalpies are in their standard states, their heat contents by convention are zero, *i.e.*,

$$H^{o}_{Cgrphite} = H^{o}_{O2(g)} = 0.$$

Hence under these conditions,

$$\Delta H^{o}_{298K} = H^{o}_{CO2(g)} = -393.5 \text{ kJ}.$$

Thus, the standard enthalpy of any compound is the enthalpy of the reaction by which it is formed from its elements in the standard state.

HEAT OR ENTHALPY OF FORMATION

The enthalpy of formation of a compound at the given temperature may be defined as the enthalpy change when one mole of a compound (in its standard state) is formed its elements (in their standard states). It is usually denoted by ΔH^{o}_{f}.

In the above definition, the standard state of any substance is taken as its natural state at 298 K under one atmosphere pressure. Let us consider the following reactions :

$$C(graphite) + O_2(g) \rightarrow CO_2(g);\ \Delta H = \Delta H^{o}_{f} = -393.5 \text{ kJ}$$

$$C(graphite) + 2H_2(g) \rightarrow CH_4(g);\ \Delta H = \Delta H^{o}_{f} = -74.8 \text{ kJ}$$

Sometimes, the enthalpy of combustion and enthalpy of formation relate to the same reaction ; for instance, the enthalpy of formation of carbon dioxide is the same as the enthalpy of combustion of graphite.

The compound with negative enthalpy of formations called exothermic compound whereas the compound with positive enthalpy of formations known as endothermic compound. Thus, the formation of water is an exothermic compound whereas the formation of hydrogen iodide is an endothermic compound.

$$H_2(g) + 1/2O_2(g) \rightarrow H_2O(g);\ \Delta H^{o}_{f} = -245.5 \text{ kJ}$$

$$1/2H_2(g) + 1/2I_2(s) \rightarrow HI(g);\ \Delta H^{o}_{f} = +25.94 \text{ kJ}$$

The negative enthalpy of formation of a compound means that its heat content is less than that of its elements and hence it is more stable than its elements. The reason is for that the compound with a lower enthalpy (or energy) must be more stable. Therefore, the chemical bonds in H_2O are stronger than those in H_2 or O_2. On the other hand, the positive enthalpy of formation of a compound means that its enthalpy of content is more than that of its elements and hence it is less stable than its elements. Due to this reason, H – I bond is weaker than H – H bond in hydrogen or I – I bond in iodine.

The enthalpy of formation of a compound may be obtained either by direct measurement of ΔH for the reaction or from enthalpies of reactions involving the compound.

HEAT OR ENTHALPY OF COMBUSTION

The enthalpy of combustion of a substance at a given temperature may be defined as the enthalpy change accompanying the complete combustion of one mole of the substance in excess of oxygen or air at that temperature.

The words "complete combustion" in the above definition must be emphasized because it is possible to burn a compound to yield both CO and CO_2. In order to avoid this confusion. The combustion is carried out in a large excess of oxygen or air to yield carbon dioxide and water as the final products. For example, heat of combustion relates to the reaction :

$$C_2H_2(g) + 21/2O_2(g) \rightarrow 2CO_2(g) + H_2O(g);\ \Delta H^\circ = -\ 1258\ kJ$$

and not to the partial oxidation as follows :

$$C_2H_2(g) + 11/2O_2(g) \rightarrow 2CO(g) + H_2O(g);\ \Delta H^\circ = -\ 691kJ.$$

The heat of combustion depends upon the physical state of the substances involved. For example, the heat of combustion of hydrogen at 298 K is –285.9 kJ mol^{-1}.

In this case only liquid water is obtained. On the other hand if the combustion is taking place at 313 K or above, the water vapour in this case cannot exist in the liquid state but evaporates to form vapours. The enthalpy of combustion is that case will be –245.5 kJ.

$$H_2(g) + 1/2O_2(g) \rightarrow H_2O(1);\ \Delta H^\circ_{298k} = -\ 285.19\ kJ\ at\ 298\ k$$

$$H_2(g) + 1/2O_2(g) \rightarrow H_2O(g);\ \Delta H^\circ_{373k} = -\ 245.5\ kJ\ at\ 373$$

The enthalpy of combustion at 373 K is lower as compared at 298 K. The reason for this is that small quantity of energy (+ 40 .44 kJ) is used up to convert liquid water into its vapours.

$$H_2O(l) \rightarrow H_2O(g);\ \Delta H^\circ -- +\ 40.44\ kJ.$$

The enthalpy of combustion of methane is

$$CH_4(g) + 2O_2(g) \rightarrow CO_2(g) + 2H_2O(l);\ \Delta H^\circ = -890.40\ kJ.$$

Enthalpy of combustion may be used to calculate the enthalpy of formation of a compound provided the enthalpy of formation of other combustion products are known.

For example, it is possible to calculate enthalpy of formation of gaseous methane provided the enthalpies of formation of CO_2(g) and H_2O(l) are known.

$$\Delta H^o = -890.40 = \Delta H^o_{CO_2(g)} + 2\Delta H^o_{H_2O(l)} - \Delta H^o_{CH_4(g)}$$

or $\Delta H^o_{CH_4(g)} = 890.40 + \Delta H^o_{CO_2(g)} + 2\Delta H^o_{H_2O(l)}$

$= 890.40 + (-393.50) + 2(-285.83) = -74.76 \text{ kJ mol}^{-1}$.

Application of Heat Combustion Date : Main applications are :

(i) Calorific value of fuels

(ii) Food values of diets

(iii) Maximum flame temperature

(iv) Calculation of heal of formation

(v) Calculation of heat of reaction.

HEAT OR ENTHALPY OF HYDROGENATION

It may be defined as the enthalpy change when a mole of an unsaturated organic compound is fully hydrogenated. The enthalpies of hydrogenation of ethylene and benzene are represented as follows .

$CH_2 = CH_2(g) + H_2(g) \rightarrow CH_3 - CH_3(g)$;

$\Delta H^o = -132.5$ kJ

$C_6H_6(l) + 3H_2(g) \rightarrow C_6H_{12}(l)$;

$\Delta H^o = -246$ kJ

From the above thermochemical equation, it is to be noted that enthalpy of hydrogenation of benzene is not three times that of enthylene one may expect this due to the presence of three double bonds in benzene and one double bond in enthylene.

The enthalpy of hydrogenation of benzene is less be 151.5kJ mol^{-1} which is attributed to the stabilisation on benzene due to resonance. This method may sometimes be used as the characteristic of stabilisation of aromatic molecules due to resonance.

HEAT OR ENTHALPY OF NEUTRALIZATION

The enthalpy of neutralization of an acid at a given temperature may be defined as *the enthalpy change accompanying the neutralization of one gram equivalent of a base by an acid in dilute solution at that temperature.*

For example, the enthalpy of neutralization one of hydrochloric acid by sodium hydroxide in dilute solution at 298 K is – 57.32 kJ.

$$NaOH + HCl \rightarrow NaCl + H_2O$$

As neutralization is the reaction between $H^+(aq)$ form the base, one can write the above equation as :

$$H^+(aq)+Cl^-(aq)+Na^+(aq)+OH^-(aq) \rightarrow H_2O(l)+Na^+(aq)+Cl^-(aq)$$

or $$H^+(aq) + OH^-(aq) \rightarrow H_2O(l)$$

From this it follows that enthalpy of neutralization should be independent of the nature of acid and the base.

This has been found to be true in the case of neutralization of strong acids by strong bases. This is evident form the data given in Table 2.

Table 2 : Enthalpies on Neutralization of Strong Acids by Strong Bases at 298 K

Acid	*Base*	*Enthalpy of Neutralization (kJ per gram equivalent)*
HCl	NaOH	–57.32
HCl	KOH	–57.45
HCl	LiOH	–57.32
HNO_3	NaOH	–57.28

Heat or Enthalpy of Ionisation

For a weak acid or a weak base, the enthalpy of neutralisation is not the same but less than –57.32 kJ. It is because of the fact that these acids or bases are not completely ionised in solution. Some of the heat is consumed in ionising these acids and bases. The heat required to ionise the weak acid or the weak base is known as the enthalpy of ionisation.

It is possible to calculate the enthalpy of ionisation of a weak acid from the enthalpies of neutralisation of strong acid and weak acid by strong bases. For example, in the neutralisation of acetic acid by sodium hydroxide, the enthalpy of neutralisation is – 55.43 kJ mol^{-1}. The enthalpy of ionisation of acetic acid will therefore be equal to +1.89 kJ mol^{-1}. this can seen as follows :

(i) $CH_3COOH(aq) + OH^-(aq) \rightarrow CH_3COO^-(aq) + H_2O(I)$;

$\Delta H^o = - 55.43$ kJ

(ii) $H^+(aq) + OH^-(aq) \rightarrow H_2O(I)$; $\Delta H = - 57.32$ kJ

(iii) $CH_3COOH(aq) \rightarrow CH_3COO^-(aq) + H^+(aq)$;

ΔH^o Ionisation = ?

On adding Eqs. (ii) and (iii), we obtain

$$CH_3COOH(aq) + OH^-(aq) \rightarrow CH_3COO^-(aq) + H_2O(I)$$

or $\Delta H^o = -57.32 + \Delta H^o$ Ionisation

But from Eq. (i), we have

$$CH_3COOH(aq) + OH^-(aq) \rightarrow CH_3COO^-(aq) + H_2O(l)$$

or $\Delta H^o = -55.43$ kJ mol^{-1}

Hence ΔH^o Ionisation = –55.43 + 57.32 = + 1.89 kJ mol.

PHYSICAL STATES OF REACTANTS AND PRODUCTS

The physical states of reactants or products have been found to influence the magnitude of heat evolved or absorbed in the reaction.

Therefore, these are indicated by symbols (g), (*l*) and (s) placed after the chemical formulae. These stand for gaseous, liquid and solid states respectively. However, sometimes confusion arises in representing the physical state of water formed as a chemical reaction. This can be understood by the following reaction at 25°C :

$$H_2(g) + \frac{1}{2}O_2(g) \rightarrow H_2O(l) + 286.2 \text{ kJ}$$

In the above reaction, since experiment is carried out at 25°C, water is obtained in the liquid state.

If the above reaction is carried out at a temperature a little above 100°C under atmospheric pressure, the water produced will remain in the vapour state. The heat evolved in the reaction is 245.5 kJ. The thermochemical equation will now be written as follows :

$$H_2(g) + \frac{1}{2}O_2(g) \rightarrow H_2O(l) + 245.5 \text{ kJ.}$$

Allotropic Forms of Substances

As different allotropic forms of a substance are associated with different amounts of energy, it means that in case of elements showing allotropy, the name of the allotropic form must be mentioned in thermochemical equation. For example, in a thermochemical equation involving sulphur, we must mention the allotropic forms, rhombic, monoclinic, etc. Thus, in a thermochemical equation, rhombic and monoclinic varieties of sulphur are represented as S_R (rhombic) and SM (monoclinic) respectively.

Reaction in Solutions

When a reaction takes place in a solution, the solvent must be indicated. For example, the symbol (aq) placed after the formula of a substance, reveals that substance is present in large excess of water to ensure that there occurs no heat change on adding further quantity of water.

Heat Changes Involved in Reactions

Two different conventions are used to represent the heat changes of the chemical processes :

(a) *Old Convention :* According to this convention, the heat evolved or adsorbed in a chemical reaction is represented as a part of the equation on the product side. A positive sign (+) is used if heat is given out and negative sign (–) is used if heat is absorbed in a chemical reaction. Thus,

$$A + B \rightarrow C + D + x \text{ kJ}$$

$$A + B \rightarrow C + D - x \text{ kJ}$$

This convention is out of date these days.

(b) *Modern Convention :* This convention is based on thermodynamics. The heat evolved or absorbed in a process may be represented by ΔH, the difference in the total heat contents of the products and reactants. It is not written as a part of equation but is written separately. If heat is absorbed in a chemical reaction, + ΔH is employed and if heat is evolved. Then –ΔH is employed. For example, the burning of carbon in oxygen to form carbon dioxide was previous represented as

$$C_{graphite} + O_2(g) \rightarrow CO_2(g) + 393.5 \text{ kJ}$$

At present it is represented as (Old convention)

$$C_{graphite} + O_2(g) \rightarrow CO_2(g);$$
$$\Delta H = -393.5 \text{ kJ}$$

Temperature and Pressure of Substances

The heat absorbed or evolved in a chemical reaction has been found to depend upon the temperature and pressure at which the reaction is being carried out. These two quantities are, therefore, also in the thermochemical equation.

If a reaction is carried out under standard conditions *i.e.*, 25°C or 298 K and 1 atm pressure, the heat change is indicated as ΔH^o. The superscript' indicates that pressure is 1 atmosphere and reaction in the standard state.

HEAT OR ENTHALPY OF SOLUTION

When a solution is dissolved in a solvent, heat may be either liberated or absorbed. If the heat is absorbed the solution becomes cooler and ΔH is given a positive sign. On the other hand if the heat is evolved, the solution gets warmer and ΔH is given a negative sign.

The enthalpy change per mole of a solute dissolved varies to a considerable extent with dilution and to some extent with the initial temperature of the solvent. Therefore, it becomes essential to express the enthalpy change with reference to the concentration of the solution. "The enthalpy of solution may be defined as the enthalpy change when one mole of the solute is dissolved in a specified amount of solvent."

The enthalpy of solution can be explained by considering the dissolution of HCl (g) in water at 298 K:

$$HCl\ (g) + 25H_2O(l) \rightarrow HCl.25\ H_2O;\ \Delta H^o = -69.5 \text{ kJ}$$

$$HCl\ (g) + 40H_2O(l) \rightarrow HCl.40\ H_2O;\ \Delta H^o = -73.18 \text{ kJ}$$

$$HCl\ (g) + 200H_2O(l) \rightarrow HCl.200H_2O;\ \Delta H^o = -74.40 \text{ kJ}$$

$$HCl\ (g) + \infty\ H_2O(l) \rightarrow HCl.(aq);\ \Delta H^o = -75.17 \text{ kJ}$$

From the above equation it means that when one mole of HCl is dissolved in 25 moles of water, 69.5 kJ of heat are evolved. But if the solution is further diluted to such an extent that further dilution would cause no additional heat change, then the equation may be written as :

$$HCl\ (g) + \infty\ H_2O(l) \rightarrow HCl.(aq);\ \Delta H^{\circ} = -75.17\ kJ\ mol^{-1}$$

where aq. represents a large volume of water. The value of ΔH° correspondents to this reaction is called the enthalpy of solution at infinite dilution. It is the maximum quantity of heat released in the formation of a solution of one mole of HCl(g) in excess of water.

For greater accuracy, two types of heat of solution have been recognised. There are :

(a) *Integral Heat of Enthalpy of Solution* : Integral heat of solution at a particular concentration is the enthalpy change when 1 mole of solute is dissolved in pure solvent to produce a solution of the given concentration. Thus, if 1 mole of solute is dissolved in 500g of water the heat change gives the value of integral heat of solution at the concentration of 2 molal.

While recording integral enthalpies of solution, the usual practice is to state the amount of solvent in which one mole of solute is dissolved. For example :

$$HCl\ (g) + 10H_2O(l) \rightarrow HCl.(10\ H_2O);\ \Delta H^{\circ} = -69.5\ kJ.$$

indicates that when one mole of hydrogen chloride gas is dissolved in 10 moles of water, there is an evolutions 65.5 kJ of heat energy.

According to Hess's law, the integral enthalpy of solution is equal to the difference between integral enthalpies of the solution at the two concentrations. For example, if to a solution of 1 mole of hydrogen chloride in 40 moles of water, enough water is added such that

$$HCl.40\ H_2O + aq \rightarrow HCl.(aq)$$

the integral enthalpy change can be calculated as follows :

$$HCl(g) + aq \rightarrow HCl\ ; \quad \Delta H^{\circ} = -\ 75.17\ kJ$$

$$HCl(g)\ 40\ H_2O \rightarrow HCl.\ 40H_2O; \quad \Delta H^{\circ} = -\ 73.18\ kJ$$

On subtracting, we have

$$HCl.40\ H_2O + aq \rightarrow HCl.(aq); \quad \Delta H^{\circ} = -\ 1.99\ kJ$$

If ΔH is plotted against number of solute per 1000 g solvent, the resultant curve is shown in Fig.3.8. Initially, ΔH increases almost linearly with m but later the increase is not as fast as m. Finally, it reaches a constant value when the solution becomes saturated with respect to solute.

If the observers value of enthalpy change (ΔH) is divided by the amount of solute (m) that is dissolved to form a solution of a particular concentration, we get the quantity $\Delta H/m$, which is nothing but the integral enthalpy of solution at the given concentration.

Integral enthalpies of solutions, enthalpies of dilutions and enthalpies of reactions in solutions can be evaluated from the enthalpies of formation in solution. In these calculations, the enthalpy of formation of water is neglected if the same number of moles of water appears on both sides of the balanced thermochemical equation.

(b) *Differential Heat or Enthalpy of Solution :* The differential heat of solution, $DH_{D.s}$ may be defined as the heat of solution of a mole of solute in such a large quantity of solution of known concentration that addition of 1 more mole of solution does not change the concentration appreciably. If on dissolving dm mole of the solute in a finite amount of the solution the heat change is dq, then differential heat of solution will be

$$\Delta H_{D.s} = \delta q/\delta m$$

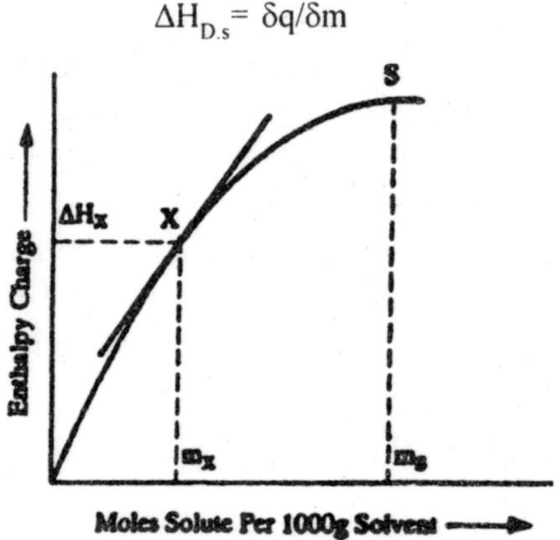

Fig. 3 : Variation of enthalpy change when the amount m of solute is dissolved in 1 kg.

From the shape of the curve of Fig. 3, it can be seen that the value of the slope will depend upon the concentration of the solution. Thus, the differential enthalpy of a solution besides depending to T and p, will also depend on the amounts of solvent n_1 and solute n_2 that are present in the given solution *i.e.*,

$$d(\Delta H)/dm = f(T, p\ n_1, n_2)$$

Thus, for this reason the concentration of the solution is mentioned while defining the differential enthalpy of solution. The following conclusions may be drawn from the shape of the curve between ΔH and m as shown in Fig. 3.

(i) For smaller value of m, the curve is almost linear; thus its slope will have a constant value and will be equal to DH/m. As the latter represents the integral heat of solution, it means that the differential and integral heats of solution are essentially equal for very dilute solutions.

(ii) For higher values on m, the curve is not linear. The reason is that ΔH does not increase as fast as m; thus the slope of the curve decreases as the value of m increases. In other words, the differential heat of solution decreases as the concentration of the solution increases, and becomes zero when the solution gets saturated.

The differential enthalpy of solution can be obtained by plotting ΔH of the solution at various concentrations against the number of moles of solvent associated with a definite quantity of solute, and finding the slope of the curve at a point corresponding to any particular concentration. The value of the slope at a given concentration gives the differential enthalpy of solution at that concentration.

The differential heats of solution and dilution are essentially the partial molar heat of solute and solvent, respectively. Consider a solution having n_1 moles of solvent and n_2 moles of solute. In general,

$$\Delta H = f(T, p\ n_1, n_2)$$

The differential of ΔH is given as follows :

$$d(\Delta H) = \left(\frac{\partial(\Delta H)}{\partial T}\right)_{P, n_1, n_2} dT + \left(\frac{\partial(\Delta H)}{\partial P}\right)_{T, n_1, n_2} dP$$

$$+ \left(\frac{\partial(\Delta H)}{\partial n_1}\right)_{T, P, n_2} dn_1 + \left(\frac{\partial(\Delta H)}{\partial n_2}\right)_{T, P, n_1} dn_2$$

At constant temperature and pressure, the above equation becomes as follows :

$$d(\Delta H) = \left(\frac{\partial(\Delta H)}{\partial n_1}\right)_{T, P, n_2} dn_1 + \left(\frac{\partial(\Delta H)}{\partial n_2}\right)_{T, P, n_1} dn_2$$

or $d(\Delta H) = \left(\Delta \overline{H}_1\right) dn_1 + \left(\Delta \overline{H}_2\right) dn_2$

where ΔH_1 and ΔH_2 are termed as partial molar values of heat of solvent and solvent and solute, respectively. It can be seen that ΔH_2 is equal to the slope d ($\Delta H/dn_2$) of a varve in which the property ΔH is plotted as a function of n_2 for a series of solutions in which n_1 is kept constant. When it is compared with the definition given at the beginning it shows that the differential heat of solution is really the partial molar heat of solution of solute, *i.e.*, ΔH_2. Similarly, the differential heat of dilution is the corresponding partial molar property of the solvent, *i.e.*, $\Delta \overline{H}_1$.

Direct experimental measurement of differential heat of solution is not possible, it can, however, be determined indirectly from the knowledge of integral heat of solution

$$\Delta H_{D.S} = d\ (m\ \Delta H_{I.S})/dm$$

where $m\Delta H_{I.S}$ is the integral heat of solution of the concentration m molal.

As it is obvious from the definitions, the integral and differential heats of solutions approach each other with decreasing the concentration and become idential when the concentration is zero.

Enthalpy of Dilution : If it known than t the enthalpy of solution per mole of a solute gets varied with the concentration of the solution. Therefore, it follows that if a given solution is diluted on the addition of solvent, there occurs noticeable change of enthalpy. This is termed as the enthalpy of dilution.

Enthalpy of dilution may be defined as the change of enthalpy when a solution having 1 mole of solute is diluted from one concentration to another.

Heat or Enthalpy of Hydration

The heat or enthalpy of hydration of salt is the change in heat content ΔH when one mole of an anhydrous substance combines with requisite number of moles of water to form the hydrate. It is obtained from the enthalpies of solutions of hydrous and anhydrous substances. Enthalpy of hydration of copper sulphate can be calculated from the following thermochemical equations :

(i) $CuSO_4.5H_2O + aq \rightarrow CuSO_4\ (aq);\ \Delta H^o = 11.72$ kJ

(ii) $CuSO_4(s) + 5H_2O \rightarrow CuSO_45H_2O\ (s);\ \Delta H^o = x$ kJ

(iii) $CuSO_4(s) + 5H_2O\ (l) + aq \rightarrow CuSO_4\ (aq);\ \Delta H^o = -66.5$ kJ

From the above equations, we get

$CuSO_4$ (s) + $5H_2O$ (l) → $CuSO_4$ $5H_2O$(s); $\Delta H^o = -78.22$ kJ.

Heat or Enthalpy of Transition

It may be defined as the enthalpy change when one mole of one allotropic form changes to another. It includes transition from solid to liquid (fusion), liquid to vapour (vaporisation), solid to vapour (sublimation) and change from one crystalline form to another crystalline form (polymorphic transition). For example, in the transition of sulphur (rhombic) to sulphur (monoclinic), The enthalpy change is 13.14 kJ.

$S_{rohmbic} \rightarrow S_{monoclinic}$; $\Delta H^o = 13.14$ kJ

Another example is

$C_{graphite} \rightarrow C_{diamond}$; $\Delta H^o = 1.90$ kJ

The enthalpy of transition in the above case can be obtained from the enthalpies of combustion of $C_{graphite}$ and $C_{diamond}$.

(i) $C_{graphite} + O_2\ (g) \rightarrow CO_2\ (g)$; $\Delta H^o = -393.51$ kJ

(ii) $C_{diamond} + O_2\ (g) \rightarrow CO_2\ (g)$: $\Delta H^o = -395.41$ kJ

On subtracting (ii) from (i), we get

$C_{graphite} \rightarrow C_{diamond}$; $\Delta H^o = 1.90$ kJ

Heat or Enthalpy of Precipitation

It may be defined at the enthalpy change when one mole of a precipitate is formed upon mixing dilute solutions of relevant electrolytes. Some examples are :

$BaCl_2$ (aq) + Na_2SO (aq) → $BaSO_4$(s) + 2NaCl (aq);

$\Delta H^o = 19.5$ kJ

As the reactants are strong electrolytes, the reaction is essentially ionic in nature and may be written as follows :

Ba^{2+} (aq) $2Cl^-$(aq) + $2Na^+$(aq) + SO_4^{2-}(aq)

→ $BaSO_4$(s) + $2Na^+$ (aq) + $2Cl^-$(aq), $\Delta H = -19.50$ kJ

As Na^+(aq) and Cl^-(aq) ions are common on both sides, the reaction, in fact, involves precipitation of $BaSO_4$(s), *i.e.*,

Ba^{2+} (aq) + SO_4^{2-}(aq) → $BaSO_4$(s); $\Delta H = -19.50$ kJ

Another example is as follows :

$AgNO_3$ (aq) NaBr (aq) → AgBr(s) + $2NaNO_3$ (aq);

$\Delta H^o = -17.78$ kJ

Enthalpy of Fusion

The enthalpy of fusion of a substance may be defined as the enthalpy change when 1 mole of the substance is converted from the solid to the liquid phase at its melting point. When 1 mole of ice melts at 273 K, the enthalpy change is approximately 6.6 kJ. This result can be put in the equation form as :

$$H_2O(s) \rightarrow H_2O(l), \qquad \Delta H = 6.0 \text{ kJ at } 273 \text{ K}$$

The enthalpies of fusion of molecular solids (O_2, CCl_4. H_2S, etc.,) are low where as the enthalpies of fusion of ionic solids (NaCI, $MgCl_2$) are large. Thus one may get some idea from the value of fusion, about the nature of the solid and the magnitude of forces acting between the particles constituting the solid.

ENTHALPY OF VAPORIZATION

The enthalpy of vaporization $\Delta(H_V)$ of a liquid may be defined as the enthalpy change accompanying the conversion of 1 mole of the substances form the liquid to the vapour phase at its boiling point. In order to cover liquid into its vapours the substance will absorb energy from the surroundings and thus ΔH will have + ve sign. For example, if water is heated at its boiling point, 373 K, it is converted into vapours by absorbing 40.6 kJ mol^{-1} energy from the surroundings. This means ΔH_V of water is 40.6 kJ mol^{-1}.

$$H_2O(l) \rightarrow H_2O(g); \qquad \Delta H_V = +40.6 \text{ kJ mol}^{-1}$$

Similarly, ΔH_V for hydrogen and oxygen are 0.904 and 6.7 kJ Mol-1 respectively. If we compare these values with water we realise that the interaction between water molecules is much stronger than that between hydrogen molecule sin the liquid state but less stronger than that between oxygen molecules in the liquid state. This is quite true because in water there is extensive hydrogen bonding whereas in oxygen and hydrogen there are only weak van der waals forces, The enthalpy of fusion is very small as compared to the enthalpy of vaporization. The reason for this is that attractive forces in solid and liquid state are much similar (intermolecular distances in solids and liquids are roughly the same) whereas in the gaseous state the attractive forces are much weaker.

Enthalpy of Sublimation

The enthalpy of sublimation (ΔH_{sub}) of a substance may be defined as the enthalpy change accompanying the conversion to 1 mole of solid directly into vapour phase at a given temperature below the melting point of the former. Generally, molecular solids like iodine and naphthalene sublime readily. For example, the enthalpy of sublimation of iodine is 62.39 kJ mol^{-1}.

According to Hess's law, the enthalpy of sublimation is equal to sum of enthalpy of fusion (ΔH_f) and enthalpy of vaporization (ΔH_V), *i.e.*,

$$\Delta H_{sub} = \Delta H_f + H_v$$

The above equation is very useful because it can be employed to calculate the enthalpy of sublimation of the enthalpy of fusion and enthalpy of vaporization are known.

Enthalpy of Atomization

This is defined as the enthalpy change required to convert one mole of an element from its normal state at 298 K and one atmosphere pressure into free atoms.

Enthalpies of atomization, are difficult to measure (they are usually obtained from spectroscopic data), but the are useful in relation to bond energies. For example :

$$H_2.g = 2H_{atoms} \qquad \Delta H = 433 \text{ kJ} \qquad ...(1)$$

$$O_2.g = 2O_{atoms} \qquad \Delta H = 495 \text{ kJ} \qquad ...(2)$$

$$H_{2g} + 1/2O_{2.g} = H_2O_g \qquad \Delta H = -246 \text{ KJ} \qquad ...(3)$$

Then (3) – (1) – 1/2 × (2) gives :

$$2H_{atoms} + O_{atoms} = H_2O \quad \Delta H = -926 \text{ kJ}$$

This energy change corresponds to the formation of two oxygen-hydrogen bonds, so that the average bond energy of an oxygen-hydrogen bond in water is 463 kJ.

BOND ENTHALPIES OR ENERGIES

The bond energy of a particular type of bond in a molecule may be defined as the amount of energy required to dissociate or break one mole of that type present in the compound and separate the resulting atoms or radicals from one another. For example, the bond energy of the H – H bond in H_2 is 433 kJ mol^{-1}. It means that 433 kJ of energy is required to break or dissociate the H – H bond in one mole of hydrogen molecules. Similarly, the bond energies of I – I in I_2 and H – I in HI are 151 kJ mol^{-1} and 229 kJ mol^{-1} respectively.

In case of diatomic molecules the bond energy is a definite quantity as there is only one type of bonds present in these molecules. However, the bond energy becomes a much less definite quantity in case of polyatomic molecules. This is to be expected since in a polyatomic molecule, the energy necessary to rupture a given bond will depend to some extend on the nature of remainder molecule. For example, the energy required to rupture the O – H bond in alcohol would not be expected to the same as the energy needed to rupture same bond in an ionic compound. This is obvious from the fact that alcohols are much less acidic than organic aids. Therefore, it follows that bond energy for a particular bond in polyatomic molecules will vary from compound to compound. However, if there are similar types of bonds present in a ployatomic molecule, the bond energy is equal to the average energy required to break a particular bond in one mole of that compound. For example, C – H bond energy in CH_4 can be calculated as follows :

When methane is formed it elements carbon and hydrogen, 1648.7 kJ of heat is liberated. It means 1648.7 kJ of energy must be supplied to one mole of methane to break four identical C – H bonds, resulting the formation of atomic carbon and atomic hydrogen in the gaseous state. It means that the average C – H bond energy is 1648.7/4 or 412.175 kJ mol^{-1}. This is also the value of heat of formation of C – H bond,

From the above discussion, it is evident that the bond energy or enthalpy should be distinguished carefully from the term bond dissociation energy of enthalpy.

Bond Enthalpy or Energy

Bond enthalpy of a given bond is defined as the average of enthalpies or energy required to dissociate the said bond present in different gaseous compounds into free atoms or radicals in the gaseous state.

Bond Dissociation Enthalpy or Energy

Bond dissociation enthalpy or energy may be defined as the enthalpy or energy required to dissociate a given bond of some specific compound.

The distinction between bond enthalpy and bond dissociation enthalpy may be better understood by a simple example, say of the O – H bond. The enthalpy of dissociation of the O – H bond depends upon the nature of molecular species from which the H atom is getting separated. For example, in the water molecule

$$H_2O(g) \rightarrow H(g) + OH(g)\ ;\ \Delta H^\circ = 501.87 \text{ kJ noet}$$
$$OH(g) \rightarrow O(g) + H(g)\ ;\quad \Delta H^\circ = 423.38 \text{ kJ noet}$$

The bond enthalpy of O – H is defined as the average of these values, *i.e.*,

$$\text{Bond enthalpy} = \frac{501.87 + 423.38}{2} = 462.625 \text{ kJ noet}$$

Hence the bond enthalpy given for any particular pair of atom is the average value of the dissociation enthalpies of the bond for a number of molecules in which the pair of atoms appear.

The bond dissociation energies for the stepwise dissociation of CH_4 are as follows :

$$CH_4(g) \rightarrow CH_3(g) + H(g);\ \Delta H^\circ = 427.0 \text{ kJ}$$
$$CH_3(g) \rightarrow CH_2(g) + H(g);\ \Delta H^\circ = 418.4 \text{ kJ}$$
$$CH_2(g) \rightarrow CH(g) + H(g);\ \Delta H^\circ = 460.2 \text{ kJ}$$
$$CH(g) \rightarrow C(g) + H(g);\ \Delta H^\circ = 343.1 \text{ kJ}$$

The energy for each step is different due to the fact that in each case a different dissociating fragment is involved. The C – H bond enthalpy would be one-fourth of the enthalpy required to break a mole of CH_4 and is 412 kJ.

CALCULATION OF BOND ENTHALPY OR ENERGY

It is possible to obtain bond energies from data on heats of combustion and heats of dissociation. Let us illustrate it by calculating the bond enthalpy or energy of C – H. We know,

$$CH_4(g) \rightarrow C(g) + 4H(g)\ \in_{C-H} = \frac{\Delta H}{4}$$

The ΔH and hence $\in_{C-H}$ for this reaction can be obtained by adding the following equations :

$$CH_4(g) + 2O_2(g) \rightarrow CO_2(g) + 2H_2O(l) \qquad \Delta H^\circ = -890.36\text{kJ}$$
$$CO_2(g) \rightarrow C\ (\text{graphite}) + O_2(g) \qquad \Delta H^\circ = 393.51 \text{ kJ}$$
$$2HO_2(l) \rightarrow 2H_2(g) + O_2(g) \qquad \Delta H^\circ = 571.70 \text{ kJ}$$
$$2H_2(g) \rightarrow 4H(g) \qquad \Delta H^\circ = 871.86 \text{ kJ}$$
$$C\ (\text{graphite}) \rightarrow C(g) \qquad \Delta H^\circ = 716.68 \text{ kJ}$$
$$CH_4(g) \rightarrow C(g) + 4H(g) \qquad \Delta H^\circ = 1663.39 \text{ kJ}$$

$$\text{Thus, at } 298.15\text{o k } \in_{C-H} = \frac{1663.39}{4} = 415.85 \text{ kJ}$$

This value of the C – H bond energy does not correspond to the dissociation energy of the carbon-hydrogen bond in methane, which is 426.77 kJ mol^{-1} and refers to the equation

$$CH_4(g) \rightarrow CH_3(g) + H(g)$$

The data on bond energies are used to calculate the approximate heat of formation of a compound of known structure by adding the appropriate bond energies. Wherever direct experimental data are not available, approximate values of enthalpies of reaction can also be obtained by using bond enthalpy data.

Estimation of the Enthalpy of Formation from Bond Enthalpies

Let us calculate the heat of formation of $Se_2CI_2(g)$ from bond enthalpies by involving the following steps ; as the bond energy refers to the dissociation of Se – Se CI gas into the gaseous atoms, the enthalpy change for the formation of this gaseous molecule should be as follows :

$$2Se(g) + 2CI(g) \rightarrow Se_2CI_2(g)$$

$$\Delta H = -[\in Se - Se + 2 \in Se - CI] = -694.54 \text{ kJ mol}^{-1}.$$

However to estimate the heat of formation it becomes necessary to add two reactions in the above equation, as by definition the heat of formation refers to the elements in their standard states. Therefore, the following enthalpy changes are introduced to convert the elements from their standard states to the gaseous atoms at 298°K

$$CI_2(g) \rightarrow 2CI(g) \; \Delta H = 243.30 \text{ kJ mol}^{-1}$$

$$2 \text{ Se (hexagonal)} \rightarrow 2 \text{ Se}(g) \quad \Delta H = 2 \times 2 \times 202.5 \text{ kJ mol}^{-1}$$

By adding these in the preceding expression, we obtain

$$2 \text{ Se (hexagonal)} + CI_2(g) \rightarrow Se_2CI_2(g);$$

$$\Delta H_f = -46.22 \text{ kJ mol}^{-1}$$

Estimation of the Heat of Reaction from Bond Enthalpies or Energies

Let us illustrate this by calculating the enthalpy change for the following equation from the bond energy data :

$$C_2H_4(g) + HCI(g) \rightarrow C_2H_5CI(g)$$

The enthalpy change for the above reaction can be calculated as follows :

ΔH = Energy required to break reactants into + Energy released to form products from the gaseous atoms gaseous atoms

$= [4\,\epsilon_{C-H} + \epsilon_{C=C} + \epsilon_{H-Cl}] + [-5\,\epsilon_{C-H} - \epsilon_{C-C} - \epsilon_{C-Cl}]$

$= (\epsilon_{C=C} + H - Cl) - (\epsilon_{C-H} + \epsilon_{C-C} + \epsilon_{C-Cl})$

On substituting the bond enthalpy values in the above equation, we get

$\Delta H = (615.05\text{kJ} + 431.79\text{kJ}) - (413.38\text{kJ} + 347.69\text{kJ} + 328.44\text{kJ})$

$= -42.67\text{kJ}$

Bond Energies and Resonance

There occurs a fair agreement between the calculated values of heat of formation obtained from bond enthalpies and any other method. However, there occurs large deviations in case of compounds having alternate single and double bonds. For example, the following reaction.

$$\underset{\text{Benzene}}{C_6H_6(g)} \rightarrow 6C(g) + 6H(g)$$

will require enthalpy of 5368.5 kJ on the basis of bond energies ($3\,\epsilon_{C=C} + 3\epsilon_{C-C} + 6\epsilon_{C-H} = 3 \times 347.69$ kJ $+ 3 \times 615.05$ kJ $+ 6 \times 413.38$kJ = 5368.5 kJ), while the experimental value is 5351.1 kJ. This amounts to the fact that benzene is more stable by 166.6 kJ mol^{-1}. This is attributed to resonance, that is, in benzene there occurs no localization of single and double bonds, but the molecule is resonating from one extreme to another.

Factors: The value of the bond energy depends upon the following factors :

(a) *Length of the bond :* In general, it is found that greater the bond length between the two nuclei of the two atoms forming the bonds, the smaller is the bond energy.

Type of bond	H – H	Cl – Cl	I – I
Bond length (i)	0.74	1.25	2.7
Bond energy (kJ/mol)	433.1	330.5	151.2

(b) *Bond polarity :* Higher is the difference in electronegativity between the combining atoms, greater is the strength of the dipole along the bond and higher is the bond energy.

(c) *Unsaturation* : In general it is observed that with the increase in the number of covalent bonds between the atoms, there occurs an increase in the bond energy but the contribution of each additional bond goes on decreasing.

Bond	*Energy in kJ/mole*
C – C	347.3
C = C	615.0
C = C	811.7

(d) *Resonance effect* : If the bonds in the molecule do not remain fixed but the molecule possesses resonance, the bond energy per bond depends upon the amount of resonance.

(e) *Overlap of Orbitals* : The bond energy depends upon the s character which is determined by extent of overlap of orbitals of the combining atoms.

SPECIFIC HEAT OF SOLIDS

The specific heat of a substance is defined as the amount of heat required to raise the temperature of 1 gm of the substance through 1°C. The specific heat of a solid increases with rise of temperature and tends to zero as the absolute zero of temperature is approached. For example, diamond (carbon) has a specific heat of 0.095 at 0°C, 0.19 at 100°C and 0.459 at 1000°C. The variation of specific heat of solids with temperature has been explained by a number of scientists. Some of the main attempts are outlined below :

Dulong and Petit's Law

The end of eighteenth century has seen great activity in what we will now call theoretical chemistry. *Lavoisier* and *Laplace* had determined the specific heats of a number of metals with considerable accuracy. In 1819, Dulong and Petit, two French scientists, pointed out that, using the accepted atomic weights for many metals and multiplying them by the specific heats, an approximately constant figure was obtained for each metal. The average value of this constant is 6.4. Thus, Dulong and Petit enunciated their law such as :

"At room temperature, the product of the specific heat and the atomic weight of a solid is a constant/or all solid elements, and is approximately equal to 6.4."

or Atomic weight × Specific Heat = –6.4 (approximately).

The product (atomic weight x specific heat) is termed as atomic heat. Hence, the Dulong and Petit's law may be stated in the alternative form :

"The atomic heats of solid elements, at ordinary temperatures, are constant and approximately equal to 6.4

Successes : When this law was applied to different solid substances, it was found that the atomic heat capacities are almost constant at about 6.2 ± 0.4 cal, per degree in spite of the regular increase in the atomic weights from 7 to 200.

Exceptions : The various exceptions to Dulong and Petit's law are as follows:

(i) According to Dulong and Petit's law, the atomic heats of all solids must be about 6.4. Actually this is not true. For example, the elements beryllium, boron, carbon and silicon have exceptionally low atomic heats: the values at ordinary temperatures being 3.5, 2-5, 1-35 (diamond) and 4.7 cal per degree respectively.

(ii) As the atomic weight of an element is constant, the atomic heat must also be constant, *i.e.*, its value should be same at all temperatures. This is contrary to the observed fact.

For example, diamond shows variation of heat capacity with temperature. This is evident from below :

Temperature	–50°C	107°C	185-5°C	260°C	615°C	808°C
Atomic Heat	076	1.135	2.11	3.28	5.33	5.44

In the case of silicon, the specific heat increases rapidly with temperature and reaches a fairly constant value above 200°C. The variation in metals like lead and tin is very small. From the above results, one can draw a very important conclusion that the Dulong and Petit's law would apply to all substances provided the temperature is high enough. However, this view is not correct because the heat capacities of those elements which appear to obey the law at ordinary temperatures increases with increasing temperatures. In some cases these values become as high as 9 cal at their melting points. Thus one can safely say that

"There is no one set of conditions under which the Dulong and Petit's law is applicable to alt elements."

(iii) Measurement of specific heats at low temperatures indicates that atomic heat decreases slowly with fall of temperature and below a certain temperature, characteristic of each element, the heat capacity decreases rapidly, tending finally to the zero value at the absolute zero of temperature. This type of change is not permitted by the law of Dulong and Petit. Thus, we can say that :

"The law of Dulong and Petit\ is entirely accidental and the atomic heat capacity of 6 cal. per degree has a theoretical significance only."

Classical Theory

L. Boltzmann (1871) deduced the value of atomic heat of solid from the kinetic theory on the assumption that :

"An ideal solid is composed of atoms vibrating about their respective equilibrium positions but not interacting with one another in any way" By the above statement it means that in a solid, atoms are allotted fixed positions about which they simply oscillate, possessing kinetic energy of translational as well as potential energy of attraction. The energy of an oscillating atom (harmonic oscillator) is given by

$$E = \text{kinetic energy} + \text{potential energy}$$

$$= \frac{p^2}{2m} + V_{(r)} = \frac{p^2}{2m} + \frac{1}{2} m\omega_0^2 x^2 \qquad ...(1)$$

where $V_{(r)}$ is equal to $1/2m\omega_0^2x^2$, m is the mass and w_0 is the natural frequency of the oscillator. Our next aim is to find out the total energy of the solid due to the various oscillating atoms present in it, *i.e.*, to sum up the energy of all atomic oscillators. The total energy E of a solid is given by

$$E = N \times \overline{E} \qquad ...(2)$$

where E is the mean energy of the atomic oscillator and N, the number of degrees of freedom enjoyed by each atom. From statistical mechanics, it follows that the number of atomic oscillators dN lying between E and E + de at a temperature T is proportional to $e^{-E/KT}$ *i.e.*,

$$dN \propto e^{-E/KT} \quad \text{or } dN = \text{constant} \times e^{-E/KT}$$

$$= \text{constant} \times e^{-(p2/2m + 1/2mwo2x3)} \qquad ...(3)$$

The mean energy of each oscillator will be given by

$$\overline{E} = \frac{\int_0^\infty E\, dN}{\int_0^\infty dN} \qquad ...(4)$$

Substituting the values of E and dN from equations (1) and (3) in (2), we get

$$\overline{E} = \frac{\int_p \int_x \left(\frac{p^2}{2m} + \frac{1}{2} m\omega_o{}^2 x^2 \right) e^{-p^2/2mKT} \cdot e^{-m\omega_o{}^2 x^2/2mKT} \, dp_x dx}{\int_p \int_x e^{-p^2/2mKT} \cdot e^{-m\omega_o{}^2 x^2/2mKT} \, dp_x dx}$$

or

$$\overline{E} = \frac{\int_{-\infty}^{+\infty} \frac{p^2}{2m} e^{-p^2/2mKT} \, dp_x}{\int_{-\infty}^{+\infty} e^{-p^2/2mKT} \, dp_x} + \frac{\int_{-\infty}^{+\infty} \frac{1}{2} m\omega_o{}^2 x^2 \, e^{-m\omega_o{}^2 x^2/2mKT} \, d\,dx}{\int_{-\infty}^{+\infty} e^{-m\omega_o{}^2 x^2/2mKT} \, dx} \quad ...(5)$$

Using the standard integrals

$$\int_0^{\infty} e^{-\alpha u^2} \, du = \frac{1}{2} \left(\frac{\pi}{\alpha} \right)^{1/2}$$

and

$$\int_0^{\infty} u^2 e^{-\alpha u^2} \, du = \frac{1}{4} \left(\frac{\pi}{\alpha^2} \right)^{1/2}$$

we have

$$\frac{\int_{-\infty}^{+\infty} \frac{p^2}{2m} e^{-p^2/2mKT} \, dp_x}{\int_{-\infty}^{+\infty} e^{-p^2/2mKT} \, dp_x} = \frac{2 \times \frac{1}{2m} \times \frac{1}{4} \left[\pi (2mKT)^3 \right]^{1/2}}{2 \times \frac{1}{2} \left[\pi \, (2mKT) \right]^{1/2}} = \frac{KT}{2}$$

$$\frac{\int_{-\infty}^{+\infty} \frac{1}{2} m\omega_o{}^2 x^2 / 2KT \, dx}{\int_{-\infty}^{+\infty} e^{-m\omega_o{}^2 x^2/2KT} \, dx} = \frac{2 \times \frac{1}{2} \times \frac{1}{4} \left[\pi (2mKT)^3 \right]^{1/2}}{2 \times \frac{1}{2} \left[\pi \, (2KT) \right]^{1/2}} = \frac{KT}{2}$$

Substituting these values in equation (5), we get

$$\overline{E} = \frac{KT}{2} + \frac{KT}{2} = KT \quad ...(6)$$

As an oscillator can vibrate about three mutually perpendicular directions, it means that the mean energy

$$\overline{E} = 3KT \quad ...(7)$$

If there are N atoms in a mole, the to total energy E is obtained by substituting eq. (7) in (2), we get

$$E = 3NKT = 3RT \quad ...(8)$$

Differentiating eq. (7) with respect to temperature, we get

$$\left(\frac{dE}{dT}\right)_V = 3R \text{ or } C_V = 3R \text{ because } C_V = \left(\frac{\partial E}{\partial T}\right)_V$$

or $(C_V = 3 \times 1.98 = 5.94$ cal. per gm. atom per A°)

The above value of 5.94 is nearly equal to the value given by the Dulong and Petit's law. This is the derivation of the law of Dulong and Petit from kinetic theory.

Limitation of Classical Theory

It could not explain the discrepancy of the decrease of specific heat at low temperatures which has been observed in case of all solids. The results are also not in accordance with experiment. The classical derivation and its predication of heat capacity to be independent of temperature are completely disproved.

Einstein Theory of Specific Heat

quantum theory of specific heat. Einstein explained the problem by using the conception of quantum theory. He assumed that the atoms in a solid are all independent, and that each atom acts as simple harmonic oscillator with a common frequency. Thus, the energy of each linear oscillator is given by

$$\in = nh\nu, \; n = 0, 1, 2 \ldots \ldots$$

where n is any positive integer.

The expression for the average energy of an oscillator in quantum theory is quite different from the classical theory. According to Planck's proposal, the linear oscillator vibrates only with integral energy values 0, $\in$, $2\in$, $3\in$, ... where $\in$ represents the elementary quantum, *i.e.*, $\in = h\nu$.

Einstein applied the concept of Planck's theory that atomic system can exist with only certain discrete energies. He further provided the key information about the dependence of heat capacities upon temperature. Before giving the details of Einstein, let us examine the general outline of the calculations. The heat capacity at constant volume is given by

$$C_V = \Delta E / \Delta T$$

where ΔE is the change in the total energy of a mole of substance produced by the temperature change, ΔT. We shall calculate ΔE from the expression

$$\Delta E = \Delta N\overline{\epsilon}$$

where N is the number of atoms in a crystal, and $\Delta\overline{\epsilon}$ is the change in the average energy of an atom in the crystal produced by the temperature increase ΔT. Our first objective then is to find an expression that gives us information how the energy $\overline{\epsilon}$ depends on temperature.

Let us consider (as Einstein did) a mole of monoatomic solid which consists of N mass points that can oscillate in the three mutually perpendicular directions with a vibrational frequency ν. According to Planck's quantized energy, an atom oscillating in one direction could have only one set of the energies given by

$$\epsilon_n = nh\nu,\ n = 0, 1, 2, 3, \ldots\ldots \qquad \ldots(1)$$

where h is Planck's constant and is a proportionality factor between frequency and energy, and n is an integer. Planck's constant has the value 6.6×10^{-27} ergs and the frequency ν is equal to 10^{12} sec^{-1} for many solids. Although an actual atom in a solid is a three-dimensional oscillator, yet we assume it to be a one-dimensional oscillator and we will correct it for this discrepancy later. At thermal equilibrium, the oscillators in a crystal will be distributed among the various allowed energies according to the Boltzmann distribution law which states that the number N_n of atoms with energy ϵ_n is related ti the number N_o with energy $\epsilon_o = 0$ by the expression

$$N_n = N_o\, e^{-\epsilon n/KT} = N_o\, e^{-nh\nu/KT} \qquad \ldots(2)$$

The total number of particles, N, is equal to the sum of the number in each energy state,

$$N = N_o + N_1 + N_2 + eN_3 + N_4 + \ldots$$

Substitution of the Boltzmann factor, eq. (2) gives

$$N = N_o + N_o e^{-h\nu/KT} + No\ e^{-2h\nu/KT} + \ldots$$

$$= N_o \sum_{n=o}^{\infty} e^{-nh\nu/KT}$$

We will not calculate the total energy due to the oscillation of the atoms in this one direction only. We can do this by multiplying energy ϵ_o by the number of particles which have that energy, and adding all these quantities :

$$\Delta E = \epsilon_o N_O + \epsilon_1 N_1 + \epsilon_2 N_2 + \ldots$$

$$= 0N_o + h\nu N_1 + 2h\nu N_2 + ...$$

$$= 0 + h\nu N_o e^{-h\nu/KT} + 2h\nu N_o e{-}2h\nu/kT + ...$$

$$= N_o \sum_{1}^{\infty} nh\nu\, e - nh\nu / kT$$

But $\overline{\in} = \frac{\Delta E}{N} = \frac{h\nu \sum ne^{-nh\nu/kT}}{\sum e^{-nh\nu/kT}}$ ...(3)

where $\overline{\in}$ is the average energy of a one-dimensional oscillator. In order to evaluate the value of $\overline{\in}$, the summation in eq. (3) must be done. This can be done if we make the substitution

$$y = e^{-h\nu/KT}, \text{ which gives } e^{-nh\nu/KT} = y^n$$

Then, for the series in the denominator of Eq. (3), we get

$$\sum_{n=1}^{\infty} e^{-nh\nu/kT} = \sum_{n=1}^{\infty} y^n = 1 + y^2 + y^3 + ... = \frac{1}{1-y} \quad ...(3A)$$

This last step can be verified by algebraic long division of 12 by $1-y$. We treat the series in the number at or of Eq. (3) in the same manner

$$\sum_{1}^{\infty} ne^{-nh\nu/kT} = \sum_{1}^{\infty} ny^n = y\left(1 + 2y + 3y^2 + ...\right) = \frac{y}{(1-y)^2} \quad ...(3B)$$

Substituting equation (3A) and (3B) in (3), we get

$$\overline{\in} = \frac{h\nu y}{1-y} = \frac{h\nu e^{-h\nu/kT}}{1-e^{-h\nu/kT}}$$

Multiplying numerator and denominator by $e^{h\nu/kT}$ yields

$$\overline{\in} = \frac{h\nu}{e^{h\nu/kT} - 1}$$

This is the average energy of a one-dimension oscillator. The average energy of an atom in a crystal is three times this amount since the atom vibrates in three directions.

Thus, $\overline{\in}_{atom} = \frac{3h\nu}{e^{h\nu/kT} - 1}$ ...(4)

We will not compute the heat capacity of the solid at such a high temperature so that $h\nu/kT << 1$. Under these conditions, the denominator of Eq. (4) simplifies considerably if we apply the fundamental properties of the exponential function.

$$e^{h\nu/kT} = 1 + \frac{h\nu}{kT} + \frac{1}{2}\left(\frac{h\nu}{kT}\right)^2 + \frac{1}{6}\left(\frac{h\nu}{kT}\right)^3 \ldots +$$

But $h\nu/kT << 1$, all terms after the second are very small and may be neglected in comparison with $1 + h\nu/kT$. Then, we get

$$e^{h\nu/kT} \approx 1 + \frac{h\nu}{kT} \qquad ...(5)$$

which when substituted into eq. (4) gives

$$\overline{\in} = \frac{3h\nu}{e^{h\nu/kT} - 1} = \frac{3h\nu}{h\nu/kT} = 3kT \qquad ...(6)$$

Therefore, the total energy of one mole of atoms is given by

$$\Delta E = \overline{N} \in = 3NkT \qquad [\backslash\ R = Nk] \qquad ...(7)$$

and the heat capacity is

$$C_V = \Delta E / \Delta T = 3R = 6 \text{ cal/ mole/ deg} \qquad ...(8)$$

The result (eq. 8) is very close to the value used in Dulong and Petit's law. The Einstein's theory thus explains specific heat of solids at high temperatures only. This has been found to be true for a large number of elements. It does not hold true in the case of elements where ν is sufficiently large. In these cases, a very high temperature is needed to attain the classical value. It means that is much depends upon the variable ν. The plot of C_V against T clearly shows that for low values of ν, it holds true.

It is more difficult to explain the behaviour of heat capacity at low temperatures by Einstein theory, because it is not possible to simplify equation (3) appreciably when T is very small. However, if we do simplification of equation (3) at low temperatures by making serious approximations, we observe tat the specific heat of solid decreases as the temperature falls and tends to be zero at absolute zero.

Successes of the Einstein Theory :

1. Einstein explained theoretically that the specific heat (atomic heat) increase with temperature and tends to the values given by Dulong and Petit at high temperatures. For some substances even at ordinary temperatures the condition is satisfied.
2. Specific heat decreases as the temperature falls and tends to be zero at absolute zero.

Limitations of Einstein Theory

1. Einstein theory is only approximate and its limitations are more serious at low temperatures. It was observed that in case of elements like copper, aluminium etc., the atomic heats at low temperatures decrease more rapidly than that predicted by Einstein theory.
2. Another weakness of Einstein theory is that while deducing his theory ν was used. This is obtained empirically and cannot be verified from any other independent physical data.
3. Einstein assumed that the vibrations of a particular atom must be of the same frequency. But the vibrations of a particular atom must be very complex, because it is under the field of force of other vibrating atoms.

SOME SOLVED PROBLEMS

Problem 1:

The heat of dissociation per mole of gaseous water at 18°C and 1 atm is 241750 J ; calculate its value at 68°C. Data given are

$C_P\ (H_2O) = 33.56;\ C_P\ (H_2) = 28.83;\ C_P\ (O_2) = 29.12 J\ K^{-1}\ mol^{-1}$

Solution:

The dissociation reaction is,

$H_2O\ (g) \rightarrow H_2\ (g) + 1/2\ O_2\ (g)\ \Delta H^\circ\ (291\ K) = 241750\ J$

$\Delta\ C_P = C_P\ (H_2) + 1/2\ C_P\ (O_2) - C_P\ (H_2O)$

$= (28.83 + 1/2 \times 29.12 - 33.56)\ JK^{-1}\ mol^{-1}$

$= 9.83 J\ K^{-1}\ mol^{-1}$

$\therefore\ \Delta\ C_P \times \Delta T = (9.83\ JK^{-1}\ mol^{-1}) \times (50K)$

$= 491.5\ J\ mol^{-1}$

$\Delta H^\circ\ (341K) = \Delta H^\circ\ (291K) + \Delta\ C_P\ (\Delta\ T)$

$= 241750\ J\ mol^{-1} + 491.5\ J\ mol^{-1}$

$= \mathbf{242241.5\ J\ mol^{-1}}$

For precise calculations we must express the heat capacities of reactants and products as a function of temperature, and then integrate equation (3) $C_{P,m}$ can be expressed as a power series of T in the form

$C_{P,m} = \alpha + \beta T + \gamma T^2$

where α, β, γ are constants for a given substance.

$\therefore \Delta C_P = (m\alpha_M + n\alpha_N) - (a\alpha_A + b\alpha_B) + [(m\,\beta_M + n\beta_N)$

$- (a\beta_A + b\beta_B)\, T + ... + ...$

or $\Delta C_P = \Delta\,\alpha + \Delta\,\beta T + \Delta\gamma T^2$

Equation (3) can be written as

$$d\,(\Delta\,H) = (\Delta\alpha + \Delta\beta T + \Delta\gamma T^2)\,dT \qquad ...(1)$$

On integration this equation yields

$$\int d(\Delta H) = \int (\Delta\alpha + \Delta\beta T + \Delta\gamma T^2)\,dT + \text{constant (I)}$$

$$\text{or } \Delta\,H = \Delta\,\alpha T + \frac{1}{2}\Delta\beta T^2 + \frac{1}{3}\,\Delta\gamma T^3 + \text{constant (I)}$$

If T = 0 then

$$\Delta\,H = \Delta\,H(T = 0) = \Delta H_0 = \text{constant (I)}$$

$$\therefore\ \Delta H = \Delta H_0 + \Delta\alpha T + \frac{1}{2}\,\Delta\beta T^2 + \frac{1}{3}\,\Delta\gamma T^3$$

ΔH_0 is the hypothetical heat of reaction at absolute zero. However, equation (1) can be integrated for suitable range of temperatures T_1 and T_2 to give

$$\int_{T_1}^{T_2} d(\Delta H) = \int_{T_1}^{T_2} (\Delta\alpha + \Delta\beta T + \Delta\gamma T^2)dT$$

$$\text{or}\quad \Delta H\,(T_2) - \Delta H\,(T_1) = \Delta\alpha\,(T_2 - T_1) + 1/2\,\Delta\beta(T_2^2 - T_1^2) + \frac{1}{3}\Delta\gamma\,(T_2^3 - T_1^3) \qquad ...(2)$$

Equation (2) is used to calculate precisely the enthalpy of reaction at any temperature T_2 provided its value at a temperature T_1 is known. The use of equation (1) or (2) is not restricted only to the chemical reactions . It can suitably be used for the physical transformations as well. For example, in a phase change (vaporization, fusion, sublimation etc). from a phase a to another phase β, the Kirchhoff's equation can be written as

$$d\,(\Delta H_{\alpha \to \beta}) = CP_{\alpha \to \beta}\ dT$$

or $\Delta H\,(T_2) = \Delta\,H(T_1) + \Delta C_P\,(T_2 - T_1)$

where $\Delta H = H\beta - H\alpha$; $\Delta\,C_P = C_P\,(b) - C_P\,(\alpha)$.

Problem 2:

For a reaction CO (g) + H_2O (g) $\rightarrow$ CO_2(g) + H_2 (g), ΔH° (298 K) = – 42.0kJ. The heat capacities of various species are given by

$C_{P,\,m}$ (CO)/J K^{-1} mol^{-1} = 26.8 + 7.0 × 10^{-3} T

$C_{P,\,m}$ (H_2O, g)/JK^{-1} mol^{-1} = 30.4 + 9.6 × 10^{-3} T

$C_{P,m}$ (CO_2)/ JK^{-1} mol^{-1} 26.0 – 43.5 × 10^{-3}T

$C_{P,\,m}$ (H_2)/J$K^{-1}$$mol^{-1}$ = 29.0 – 0.80 × 10^{-3} T

Calculate (a) ΔH° (1298K) (b) ΔE° (1298 K)

Solution:

ΔH° (1298 K) is given by

$$\Delta H^{\circ}\ (1298K) = \Delta H^{\circ}\ (298K) + \int_{298}^{1298} \Delta C_P dT$$

Now, $\Delta C_P = [C_{P,m}(CO_2) + C_{P,m}(H_2)] - [C_{P,m}(CO) + C_{P,m}(H_2O)]$

= (26.0 + 29.0) – (26.8 + 30.4) + (43.5 – 0.80)

– (7.0 + 9.6)] × 10^{-3}T

= – 2.2 + 26.1 × 10^{-3} T

$$\therefore \int_{298}^{1298} \Delta C_P dT = 2.2 \int_{298}^{1298} dT + 26.1 \times 10^{-3} \int_{298}^{1298} TdT$$

$$= -2.2\,[T]_{298}^{1298} + 26.1 \times 10{-}3 \times \frac{1}{2}[T]_{298}^{1298}$$

$$= -2.2\,[1298 - 298] + 26.1 \times 10{-}3 \times \frac{1}{2}[(1298)^2 - (298)^2]$$

= 18627 J = 18.627 kJ

$\therefore$ ΔH° (1298K) = – 42.0 kJ + 18.627 kJ = 23.373kJ

Δ 67 (1298 K) = ΔH° (1298K) – Δv_g RT = ΔH° (1298K)

= – **23.373 kJ**

Since Δv_g = (1 + 1) – (1 + 1) = 0 for the given reaction.

Problem 3:

When hydrogen is burnt is oxygen, 240.5 kJ heat is liberated per mole of the explosion product measured at 27°C, Calculate the maximum temperature of explosion, $C_{V,\,m}$ (H_2O, g) = 24.15 JK^{-1} mol^{-1}.

Solution:

The reaction can be represented as

$$H_2\ (g) + \frac{1}{2}O_2\ (g) \rightarrow H_2O\ (g)\ \Delta E\ (300K) = -240.5\ kJ\ mol^{-1}$$

This amount of heat will be utilized to raise the temperature of the product $H_2O(g)$. Using equation (15) we get

$$240500\ J\ mol^{-1} = \int_{300}^{T_2} (24.15 JK^{-1}mol^{-1})dT$$

$$\text{or}\ \ 240500 = 24.15\ (T_2 - 300)$$

T_2 = **10258 K**

This calculation is quite approximate, therefore, this temperature would be far from the actual value.

Problem 4:

The heat of combustion of methane is – 881.25 kJ mol⁻¹ at 298 K. Calculate the maximum theoretical flame temperature when one mole of CH_4 is completely burnt in calculated amount of air just sufficient for combustion at a constant pressure. The necessary data of heat capacities and their temperature dependence are also given.

Data given

$$CH_4(g) + 2O_2(g) \rightarrow CO_2(g) + 2H_2O(l)$$

$$\Delta H^\circ\ (298\ K) = -\ 881.25\ kJ\ mol^{-1}$$

$$H_2O\ (l) \rightarrow H_2O(g) \quad \Delta H^\circ(298\ K) = 43.60\ kJ\ mol^{-1}$$

$$C_{P,m}\ (CO_2)/JK^{-1}mol^{-1} = 26.00 + 43.5 \times 10^{-3}T - 148.3 \times 10^{-7}\ T^2$$

$$C_{P,m}\ (H_2O,g)/JK^{-1}\ mol^{-1} = 30.36 + 9.61 \times 10^{-3}\ T + 11.8 \times 10^{-7}T^2$$

$$C_{P,m}\ (N_2)/JK^{-1}\ mol^{-1} = 27.30 + 5.23 \times 10^{-3}\ T - 0.04 \times 10^{-7}T^2$$

Solution:

Let T_2 be the maximum theoretical flame temperature to which the products are raised, which include one mole $CO_2(g)$, 2 moles $H_2O(g)$, and 8 moles $N_2(g)$ (present in air associated with 2 moles O_2 used for combustion). Since water is ultimately in the gaseous form at high temperature, that value of ΔH° os required for the reaction on which H_2O (g) is a product. Thus

(i) CH_4 (g) + $2O_2$(g) → CO_2 (g) + $2H_2O$(l) $\Delta H° = -881.25$ kJ

(ii) $2H_2O$ (l) → $2H_2O$(g) $\Delta H° = +87.20$ kJ

so that, addition of (i) and (ii) gives

CH_4 (g) + $2O_2$(g) → CO_2 (g) + $2H_2O$ (g) $\Delta H° = -794.05$ kJ

Now, this heat is used to raise the temperature of products after combustion. Equation (14) gives,

$$794050 \text{ J mol}^{-1} = \int_{298}^{T_2} \Sigma n(C_P)dT \qquad ...(1)$$

Now, $\Sigma nC_{P,m} = C_{P,m}(CO_2) + 2C_{P,m}(H_2O) + 8CP,m(N_2)$

$\Sigma (nC_{P,m}) = 305.12 + 104.56 \times 10^{-3}$ T $- 125 \times 10_{-7}$ T_2

Therefore, putting the values in equation (17) and carrying out integration we have

$$794050 = \left[30.512T + \frac{1}{2} \times 104.56 \times 10^{-3} T^2 - \frac{1}{3} \times 125 \times 10^{-7} T^3\right]_{298}^{T^2}$$

This equation is solved for T_2. This approximate value of T_2 is found to be 2250 K. The experimental value is about 2150 K. Thus there is a difference of about 100° between theoretical and experimental values. This discrepancy is explained taking into account the various possible reasons.

(i) The process may not be exactly adiabatic and there is exchange of heat between the system and the surroundings.

(ii) In actual practice, excess air is used for combustion process therefore, after combusion some air (O_2 + N_2) is left. Thus the heat evolved will be used to raise the temperature of extra O_2 and N_2 (in air).

(iii) At high temperature attained in the burning of hydrocarbons the products H_2O and CO_2 may get dissociated as

$$H_2O \rightarrow H_2 + \frac{1}{2}O_2$$

$$CO_2 \rightarrow CO + \frac{1}{2}O_2$$

These reactions will involve considerable amount of heat which is not accounted for by the theoretical calculations.

(iv) The combustion may not be complete, hence along with the products some reactants may also be present, and heat evolved will be utilized to raise the temperature of unreacted species.

Problem 5:

It is estimated that an average human consumes the equivalents of 10 g glucose per hour. Estimate the power output of the brain in watts. Given that 1 W = 1 JS^{-1};

ΔH^o_f (H_2O, l) = –286 kJ mol^{-1}

ΔH^o_f (CO_2 g) = –394 kJ mol^{-1}

ΔH^o_f (*glucose, aq*) = –1260 kJ mol^{-1}

Solution:

The heat supplied by 10 g glucose can be calculated if we know its heat of combustion:

$C_6H_{12}O_6$ (aq) + $6C_2$ (g) → $6CO_2$ (g) + $6H_2O$ (l) ΔH = ?

$\Delta H = 6 \times \Delta H^o_f (CO_2) + 6 \times \Delta H^o_f (H_2O) - \Delta H^o_f (C_6H_{12}O_6, aq)$

$= [(-6 \times 394 - 6 \times 286) + 1260]$ kJ mol^{-1}

= – 2820 kJ mol^{-1}

∴ Heat supplied by 10 g glucose = $-2820 \text{ mol}^{-1} \times \dfrac{10\text{g}}{180\text{g mol}^{-1}}$

q = –156.66 kJ per hour

= – 156666 J per hour

Power output of the brain = *Heat change per second*

$= \dfrac{156666\text{J}}{60 \times 60}$ per second

= 43.5 Js^{-1} = 43.5 W

Problem 6:

Calculate the enthalpy change for the process

H_2O (l, –10°C) → H_2O (s, – 10°C)

given that :

$C_{P,m}$ (H_2O, l) = 75.4 J K^{-1} mol^{-1}

$C_{P,m}$ (H_2O,s) = 37.2 J K^{-1} mol^{-1}

H_2O (l, 0°C) → H_2O (s, 0° C), ΔH = – 6008 J mol^{-1}

Solution:

The given process may be considered to proceed in different stages as

(i) H_2O (l, –10°C) $\xrightarrow{\text{Heat}}$ H_2O (l,0°C)

$$\Delta H \text{ (i)} = \int_{263}^{273} C_{P,m}(H_2O, l)dT = 75.4 \text{ JK}^{-1} \text{ mol}^{-1} (273\text{K} - 263\text{K})$$

= 754 J mol^{-1}

(ii) H_2O (l, 0°C) → H_2O(s, 0°C)

ΔH (ii) = – 6008 J mol^{-1}

(iii) H_2O (s, 0°C) $\xrightarrow{\text{Cool}}$ H_2O (s, – 10°C)

$$\Delta H \text{ (iii)} = \int_{273}^{263} C_{P,m}(H_2O, s)dT = 37.2 \text{ KJ}^{-1} \text{ mol}^{-1} (263 \text{ K} - 273 \text{ K})$$

= – 372 J mol^{-1}

Addition of equations (i), (ii) and (iii) gives

H_2O (l, – 10°C) → H_2O (s,–10°C), ΔH

= Δ H (i) + ΔH (iii) Δ H = – 5626 J mol^{-1}

Problem 7:

The following reactions might be used to power rockets.

(i) H_2 (g) + 1/2 O_2 (g) → H_2O(g) ΔH (i) = – 242 kJ mol^{-1}

(ii) CH_3OH (l) + 3/2 O_2 (g) → CO_2 (g) + $2H_2O$ (g)

ΔH (ii) = – 640 kJ mol^{-1}

(iii) H_2 (g) + F_2 (g) → 2HF (g) ΔH (iii) = – 540 kJ mol^{-1}

Calculate the enthalpy changes for each of these reactions per kg of each second reactant.

Solution:

Moles of reactants can be calculated by using the formula n = mass/ molar mass.

$$n\ (H_2) = \frac{1\text{kg}}{0.002 \text{ kg mol}^{-1}} = 500 \text{ mol}$$

$$n(O_2) = \frac{1kg}{0.032\ kg\ mol^{-1}} = 31.25\ mol$$

$$n(CH_3OH) = \frac{1kg}{0.032\ kg\ mol^{-1}} = 31.25\ mol$$

$$n(F_2) = \frac{1kg}{0.038\ kg\ mol^{-1}} = 26.3\ mol$$

The Stoichoiometry of reaction (i) indicates that 31.25 moles O_2 will combine with 62.5 moles H_2 to give 62.5 moles H_2O as

$H_2(g) + 1/2\ O_2(g) \rightarrow H_2O(g)$, ΔH (i) = – 242 kJ mol^{-1}

$62.5\ H_2(g) + 31.25\ O_2(g) \rightarrow 62.5\ H_2O\ (g)$

$\Delta H = 62.5 \times DH$ (i) = – **15125 kJ**

The second reaction is written as

$20.8\ CH_3OH\ (l) + 31.25\ O_2\ (g) \rightarrow 20.8\ CO_2\ (g) + 41.6\ H_2O(g)$

$\Delta H = 20.8 \times \Delta H$ (ii)

= **–13312 kJ**

Third reaction

$26.3\ H_2\ (g) + 26.3\ F_2\ (g) \rightarrow 52.6\ HF\ (g)$

$\Delta H = 26.3 \times \Delta H$ (iii)

= **– 14202 kJ**

Problem 8:

An average man weighs about 70 kg and produces about 10^4 kJ of heat per day through metabolic activity. (a) Suppose that a man were an isolated system and that his heat capacity is 4.2 $JK^{-1}\ g^{-1}$. If his temperature was 37°C at a given time what would be his temperature 24 hours later?

(b) Man is actually an open system and the main mechanism of loss of heat is evaporation of water. How much water would need to be evaporated per day to maintain the body temperature at 37° C? ΔH_{vap} (water) = 2405 J g^{-1} at 37°C.

Solution:

(a) Let T_2 be temperature of the body after 24 hours. Given that C_P = 4.2 $JK^{-1}\ g^{-1}$,

$m = 70 \text{ kg} = 70 \times 10^3 \text{ g}$

$\Delta H = 10^4 \text{ kJ} = 10^7 \text{ J}, T_1 = 310\text{K}$

$$\Delta H = \int_{T_1}^{T_2} mC_P dT = mC_P(T_2 - T_1)$$

$$\therefore T_2 = \frac{\Delta H + mC_P T_1}{mC_P}$$

$$= \frac{10^7 \text{J} + (7 \times 10^4 \text{g}) \times (4.2 \text{JK}^{-1}\text{g}^{-1}) \times (310\text{K})}{(7 \times 10^4 \text{g}) \times (4.2 \text{JK}^{-1}\text{g}^{-1})}$$

$= \mathbf{344\ K = 71^\circ\ C.}$

(b) $\because$ 2405 J heat evaporate 1g H_2O

$\therefore$ 10^7 J heat will evaporate $\frac{10^7 \text{J} \times 1\text{g}H_2O}{2405\text{J}}$

$= 4158 \text{ g } H_2O$ = **4.2 kg H_2O.**

This much water should be evaporated by perspiration to maintain the body temperature.

Problem 9:

In an explosion reaction 21000 kJ *are evolved. Calculate the decrease in mass accompanying the reaction.*

Solution:

$\Delta E = \Delta mc^2$ (Einstein equation)

$\Delta E = -21000 \text{ kJ} = -21.0 \times 10^6 \text{ T} = -21.0 \times 10^6 \text{ kg m}^2\text{s}^{-2}$

Speed of light = c = $3.0 \times 10^8 \text{ ms}^{-1}$

$$\therefore \Delta m = \Delta E/c^2 = -\frac{21.0 \times 10^6 \text{kg m}^2\text{s}^{-2}}{(3.0 \times 10^8 \text{ms}^{-1})^2} = -\mathbf{2.3 \times 10^{-10}\ kg}$$

Thus 2.3×10^{-10} kg is the loss of matter which liberates as heat and is transferred to the surroundings.

Exercise. *Energy is emitted by the sun at the rate of 2 ergs per g of the sun per second. The estimated mass of the sun is 2×10^{33} tons. Estimate the decrease in tons and per cent decrease in the mass of the sun per day.*

[**Ans.** 4×10^{17} tons; 2×10^{-14}%. It is interesting to note that if the above calculations were true then after 10^{13} years the sun should vanish.]

Problem 10:

Calculate the heat of formation of acetic acid if heat of combustion is 869.0 kJ mol⁻¹. The heats of formation of $CO_2(g)$ and $H_2O(l)$ are – 395.0 kJ mol⁻¹ and – 285.0 kJ mol⁻¹ respectively.

Solution:

The data given are expressed in the form of three the mochemical equations.

(1) $C(s) + O_2(g) \rightarrow CO_2(g);$ $\Delta H_1 = -395$ kJ

(2) $H_2(g) + 1/2O_2(g) \rightarrow H_2O(l);$ $\Delta H_2 = -285$ kJ

(3) $CH_3COOH(l)+2O_2(g)\rightarrow 2CO_2(g)+2H_2O(l);$ $\Delta H_3 = -869$ kJ

The equation corresponding to the formation of acetic acid from its elements is

$C(s) + 2H_2(g) + O_2(g) \rightarrow CH_3COOH(l)$; $\Delta H = ?$

This equation may be obtained from the three thermochemical equations on multiplying equations (1) and (2) by two and adding and finally subtracting equation (3);

$$2C(s) + 2O_2(g) \rightarrow 2CO_2(g)$$

$$2H_2(g) + O_2(g) \rightarrow 2H_2O(l)$$

Adding $2C(s) + 3O_2(g) + 2H_2(g) \rightarrow 2CO_2(g) + 2H_2O(l)$

Subtract Eq. (3) $CH_3COOH(l) + 2O_2(g) \rightarrow 2CO_2(g) + 2H_2O(l)$

$$2C(s) + O_2(g) + 2H_2(g) \rightarrow CH_3COOH(l)$$

Rearranging

$$2 \times (1)2 \times \Delta H_1 = -790 \text{ kJ}$$

$$2 \times (2)2 \times \Delta H_2 = -570 \text{ kJ}$$

Adding $\Delta H' = -1360$ kJ

Subtract Eq. (3) $\Delta H_3 = -869$ kJ

Giving $\Delta H = -1360 - (869) = -$ kJ.

Problem 11:

The enthalpy of evaporation of water at 373 K is 40.67 kJ mol⁻¹. What will be the enthalpy of evaporation at 353 K and 393 K if average molar heats at constant pressure in this range for water in liquid and vapour states are 75.312 and 33.89 JK⁻¹ mol⁻¹ respectively ?

Solution:

$$H_2O(l) \rightleftharpoons H_2O(g)$$

Enthalpy change at 373 K is ΔH_1 = 40.67 kJ mol^{-1}

= 40670 J mol^{-1}

$$\Delta H_2 - \Delta H_1 = \Delta C_p(T_2 - T_1)$$

$$\Delta H_2 = \Delta H_1 + \Delta C_1(T_2 - T_1)$$

$$\Delta C_p = \Delta C_{p,\ H2O(g)} - \Delta C_{p,\ H2O(l)}$$

$$= 33.89 = 75.312 = - 41.422.$$

Enthalpy of evaporation at 353 K, ΔH

= 40670 – 41.522(353 – 373)

= 40670 + 41.421 × 20

= 41498.44 J.

Similarly enthalpy of evaporation at 393 K

= 40670 – 41.422(393 – 373)

= 39841.56 J.

Problem 12:

The heat of formation of one mole of HI from hydrogen and iodine vapour at 25°C is 8000 cal.

$$1/2H_2(g) + 1/2I_2(g) \rightarrow HI(g),\ \Delta H = - 8000\ cal.$$

Calculate the heat of formation at 10°C and also the total change in the heat capacity at constant pressure. It is given that the molar heat capacities of hydrogen, iodine and HI vapours are given by the equations :

$C_p = 6.5 + 0.0017T$	*for*	*Hydrogen(g)*
$C_p = 6.5 + 0.0031T$	*for*	*Iodine(g)*
$C_p = 6.5 + 0.0016T$	*for*	*HI(g)*

where T is the absolute temperature.

Solution:

$$\Delta C_p = C_{p(HI)} - 1/2\ [C_{p(HI)} + C_{P(12)}]$$

$$= 6.5\ \ 0.0016T - 1/2\ [6.5 + 0.0017T + 6.5 + 0.0038T]$$

$$= - 0.00115T\ cal.\ deg^{-1}.$$

Here $\Delta H_{298} = \Delta H_{283} + \int_{283}^{298} \Delta C_p \, dT$

where $T_1 = 273 + 10 = 283$ K, $T_2 = 273 + 25 = 298$ K

Also, ΔH_{298} K = – 8000 cal.

$$-8000 = \Delta H_{283} + \int_{283}^{298} (-0.00115) \; T \, dT$$

or $\Delta H_{283} = -8000 + \frac{0.00115}{2}[T_2]_{283}^{298} = -8000 + \frac{0.00115}{2}\left[(298)^2 - (283)^2\right]$

$= -7995$ cal (nearly).

Problem 13:

What are the reasons for the 'calculated maximum adiabatic flame temperature' not being exactly equal to theexperimental value ? What correction would you suggest ?

Solution:

The maximum adiabatic flame temperature is calculated using the expression

$$T_f = \frac{-\Delta H}{C_p(\text{products})} + T_0$$

In derving the above expression, it is assumed to C_ps are independent of temperature. However, this is not correct since T_f is a very large quantity (as ΔH is expressed in kJ whereas C_ps are in joule) of the order of thousands of kelvin. Besides this, the above value will be obtained if the combustion process is carried out reversibly, since only then q = ΔH. But the conditions of experiment are those of irreversible process for which q_p is expected to be less than ΔH. Moreover, it is assumed that Cp (Products) are independent of temperature. To get correct result, one should take into account the variation of heat capacities with temperature.

Problem 14:

How much heat is required to raise the temperature of 1 mole of oxygen from 303 K to 1300 K at constant pressure?

$$C_p = 6.095 + 3.253 \times 10^{3} T - 1.017 \times 10^{6} T^2.$$

Solution:

$$H_2 - H_1 = \int_{T_1}^{T_2} C_p \, dT$$

$$T_1 = 300 \text{ K}, T_2 = 100\text{K}.$$

$$H_{1300} - H_{300} = \int_{300}^{1300} \left(6.095 + 3.253 \times 10^{-3} T - 1.07 \times 10^{-6} T^2\right) dT$$

[Substituting the value of C_p]

$$H_{1300} - H_{300} = 6.095 (1300 - 300) + \frac{3.253 \times 10^{-3}}{2} \times \left[(1300)^2 - (300)^2\right]$$

$$- 1/3 \times 1.017 \times 10{-}6 \, [(1300)^3 - (300)^3]$$

$$= 6095 + 2602.4 - 735.63 = 7961.77 \text{ cal/mole.}$$

Problem 15:

Calculate ΔH at 25°C for the reaction

$$H_2O(g) \rightarrow H_2(g) + 1/2 O_2(g)$$

given that ΔH at 18°C is 241.750 kJ mol^{-1} and molar-Heat capacities expressed as Jk^{-1} $mole^{-1}$ are $C_p(H_2) = 28.83$, $C_p(O) = 29.12$ $C_p(H_2O) = 33.56$

Solution:

$\Delta C_p^{\bullet}$ = Heat capacities of products – Heats capacities of reactants

$= [28.83 + 1/2 \times 29.12] - 33.56 = 43.39 - 33.56 = 9.83 \, Jk^{-1} mole^{-1}$

$$T_2 - T_1 = (273 + 25) - (273 + 18) = 7 \text{ k}^{\cdot}$$

$$\Delta H_2 + \Delta H_1 = \Delta C_p (T_2 - T_1)$$

or $$\Delta H_2 = \Delta H_1 + \Delta C_p (T_2 - T_1) = 241.750 + 9.83 \times 7$$

$$= 310.56 \text{ kJ mol}^{-1}.$$

Problem 16:

A student made the following erroneous statement in a laboratory report on bomb calorimetry.

$$\Delta H = \Delta U + p\Delta V$$

Since the bomb calorimetry process is a constant volume one, $\Delta V = 0$ and $\Delta H = \Delta U$. Explain why this argument is incorrect

Solution:

Since, $$H = U + pV$$

we will have $dH = dU + pdV + Vdp$.

For a constant volume process, the above expression is reduced to

$$dH = dU + V\,dp$$

While writing the expression

$$\Delta H = \Delta U + p\,\Delta V$$

the student has not written the expression V dp which makes a significant contribution when the system involves the gaseous phase. Hence, his conclusion $\Delta H = \Delta U$ is incorrect.

Problem 17:

Using bond enthalpies from the following table evaluate ΔH for the isomerization of ethyl alcohol to dimethyl ether :

Bond	*Bond enthalpies at 25°C* $\Delta H . kJ\ mol^{-1}$
H – H	416
O – O	132
O – H	444
C – H	396
C – O	336
C – C	332

Solution:

We have to calculate the enthalpy change of the reaction

$$CH_3CH_2OH \rightarrow CH_3OCH_3$$

The enthalpy change of the reaction will be given as

$\Delta H = \epsilon(C-C) + 5 \times \epsilon(C-H) + \epsilon(C-O) + \epsilon(O-H) - 6 \times \epsilon(C-H) - 2 \times \epsilon(C-O)$

$= \epsilon(C-C) - \epsilon(C-H) - \epsilon(C-O) + \epsilon(O-H) = (332 - 396 - 336 + 444)\ kJ\ mol^{-1} = 44\ kJ\ mol^{-1}$.

Problem 18:

H_2 gas is mixed with air at 25°C under a pressure of 1 atmosphere and exploded in a closed vessel. The heat of the reaction

$$H_2(g) + 1/2\ O_2(g) \rightarrow H_2O(g)$$

at constant volume, $\Delta E_{298} = -\ 240.60$ kJ mol^{-1} and C_p values for H_2O vapour and N_2 in the temperature range 298 K and 3,200 K are 39.06 JK^{-1} mol^{-1} and 26.40 JK^{-1} mol^{-1} respectively. Calculate the explosion temperature under adiabatic conditions.

Solution:

For a constant volume process under adiabatic conditions,

$$\Delta E = \Delta E_{heating} + \Delta E_{298} = 0$$

$$\therefore \quad \Delta H_{heating} = -\ \Delta E_{298}$$

$$= -\int_{298}^{T_f} n \sum C_V\, dT = -\ 240.60 \text{ kJ mol}^{-1} \qquad ...(i)$$

As 2 moles of unreacted N_2 are associated with 1/2 mole of O_2, we have

$$\Sigma nC_V = C_V(H_2O, g) + 2CV\ (N_2,g)$$

$$= (39.06 + 2 \times 26.40)\ \text{Jk}^{-1}\ \text{mol}^{-1}$$

$$= 91.86\ \text{Jk}^{-1}\ \text{mol}^{-1}$$

Hence, from Eq. (i) on integrating, we get

$$(91.86\ \text{Jk}^{-1}\ \text{mol}^{-1})\ (T_f - 298) = 240{,}600\ \text{J mol}^{-1}$$

$$T_f - 298 = \frac{240600\ \text{jmol}^{-1}}{91.86\ \text{J k}^{-1}\ \text{mol}^{-1}} = 2619\ \text{k},$$

$$T_f = (2619 + 298)\ \text{k} = 2917\ \text{k}.$$

Problem 19:

Estimate the adiabatic flame temperature of a spirit lamp, provided the following data

$$C_2H_5OH(l) + 3O_2(g) \rightarrow 2CO_2(g) + 3H_2O(g)$$

$$\Delta H_{298}\ k = 1370\ \text{kJ mol}^{-1}.$$

$C_{p.m}$ *for (i)* $CO_2(g) = 36.43$ *J k^{-1} mol$_{-1}$,*

(ii) $H_2O(g) = 33.70$ *J k^{-1} mol^{-1}, and*

(iii) $N_2(g) = 294.4$ *J k^{-1} mol$_{-1}$.*

If the above reaction is carried out in air consisting of 80 % N_2 and 20% of O_2 by volume, what would be the final temperature ?

Solution:

The expression of adiabatic flame temperature is

$$T_f = \frac{-\Delta H}{C_p(\text{products})} = T_o$$

$$= \frac{\left(1370 \times 10^3 \text{ J mol}^{-1}\right)}{(2 \times 36.43 + 3 \times 33.70)\text{J k}^{-1}\text{ mol}^{-1}} + 298\text{k}$$

$$= 7875 \text{ k} + 298 \text{ k} = 8173 \text{ k}$$

If the reaction is carried out in air, the consumption of 3 mol of O_2 would leave 12 mol of N_2 from air. Hence,

$$T_f = \frac{\left(1370 \times 10^3 \text{ J mol}^{-1}\right)}{(2 \times 36.43 + 3 \times 33.70 + 12 \times 29.4)\text{J k}^{-1}\text{ mol}^{-1}} + 298\text{k}$$

$$= 2600 \text{ k} + 298 \text{ k} = 2898 \text{ k}$$

Problem 20:

Use Born-Haber cycle to determine the lattice energy of KCl(s) at 298.15 K. The following data are given at the same temperature.

Enthalpy of formation of KCl	*= – 435.9 k J mol[1]*
Ionization energy of K	*= 418.9 k J mol[1]*
Electron affinity of Cl	*= 348.6 k J mol[1]*
Dissociation enthalpy of Cl_2	*= 242.0 k J mol[1]*
Sublimation enthalpy of K	*= 89.8 k J mol[1]*

Solution:

We are given that

(1) $K(s) + 1/2\, Cl_2(g) \rightarrow KCl(s)$ $\Delta H_1 = 435.9$ kJ mol^{-1}

(2) $K(g) \rightarrow K^+(g) + e$ $\Delta H_2 = 418.9$ kJ mol^{-1}

(3) $Cl(g) + e \rightarrow Cl^-(g)$ $\Delta H_3 = -348.6$ kJ mol^{-1}

(4) $Cl_2(g) \rightarrow 2Cl(g)$ $\Delta H_4 = 242.0$ kJ mol^{-1}

(5) $K(s) \rightarrow K(g)$ $\Delta H_5 = 89.8$ kJ mol^{-1}

We have to find ΔH of the following equation

(6) $K^+(g) + Cl^-(g) \rightarrow KCl(s)$

Equation (6) can be obtained by the following manipulations.

Eq. (1) – Eq. (5) – Eq. (2) (2) – 1/2 Eq. (4) – Eq. (3)

Accordingly, we have

$$\Delta H_6 = \Delta H_1 - \Delta H_5 - \Delta H_2 - \Delta H_1 .2\ \Delta H_4 - \Delta H_3$$

$\Delta H6 = (-435.9 - 89.8 - 418.9 + 1/2 \times 242.0 + 348.6)$ kJ mol^{-1}

$= -475.0$ kJ mol^{-1}.

Problem 21:

The bond enthalpy of $H_2(g)$ is 436 kJ mol^{-1} and that of $N_2(g)$ is 941.3 kJ mol^{-1}. Calculate the average bond enthalpy of an N – H bond in ammonia ΔH_f^o (NH_3) = – 46.0 kJ mol^{-1}.

Solution:

∵ (i) $N_2(g) \rightarrow 2N(g)$; $\Delta H^o = 941.3$ kJ mol^{-1}

(ii) $H_2(g) \rightarrow 2H(g)$; $\Delta H^o = 436.0$ kJ mol^{-1}

(iii) $1/2N_2(g) + 3/2H_2(g) \rightarrow NH_3(g)$; $\Delta H^o = -46.0$kJ mol^{-1}

On multiplying equation (i) by 1/2 and equation (ii) by 3/2 and adding, we get

(iv) $1/2N_2(g) + 3/2H_2(g) \rightarrow N(g)+3H(g)$; $\Delta H^o = 1124.6$kJ mol^{-1}

On subtracting equation (iii) from (iv), we get

(v) $NH_3(g) \rightarrow N(g) + 3H(g)$, $\Delta H^o = 1170.6$ kJ mol^{-1}

As there are three N – H bonds, in NH_3, the average bond enthalpy is obtained by dividing the value ΔH^o from (v) by 3. Hence

$\Delta H_{N-H} = 1170.6/3 = 390.2$ kJ mol^{-1}

Problem 22:

0.50 f of benzoic acid was subjected to combustion in a bomb caloi imeter at 15°C when the temperature of the calorimeter system (including water) was found to rise by 0.55°C. Calculate the heat of combustion of benzoic acid : (i) at constant volume, (ii) at constant pressure. The thermal capacity of the calorimeter system including water was found to be 23.85 kJ.

Solution:

(i) Heat of Combustion at Constant Volume (ΔE)

Weight of benzoic acid, $m = 0.50$ g

Rise in temperature, $\theta = 0.55^{\circ}C$

Molecular weight of benzoic acid, $M = 122$

Thermal capacity of the calorimeter system, z $= 23.85$ kJ

$\therefore$ Heat of combustion at constant volume, ΔE $= z \times \theta \times M/m$

$$= -23.85 \times 0.55 \times \frac{122}{0.50} = -3200.7 \text{ kJ} = -3200.7 \text{ kJ}$$

(ii) Heat of Combustion at constant Pressure (ΔH)

The combustion of benzoic acid at 25°C involves the following reaction :

C_6H_5COOH (s) + $7.5O_2(g) \rightarrow 7CO_2(g) + 3H_2O(l)$

We known that

$\Delta H = \Delta E + \Delta nRT$

In the present case :

$\Delta E = -3200.7$ kJ $\Delta n = 7 - 7.5 = -2.5$ $R = 8.314 \times 10^{-3}$ kJ

$T = 273 + 25 = 298^{\circ}K$

$\therefore \quad \Delta H = -3200.7 + (-0.5) \times 8.314 \times 10^{-3} \times 298 = -3201.9$ kJ

$\therefore$ Heat of Combustion at constant pressure,

$\Delta H = -3201.9$ kJ

Problem 23:

The heats of formation of carbon dioxide and water are 395 kJ and 285 kJ respectively, and the heat of combustion of acetic acid is 869 kJ (all exothermic). Calculate the heat of formation of acetic acid.

Solution:

The date given are expressed in the form of three thermochemical equations :

$C + O_2 = CO_2$ $\Delta H_1 = -395$ kJ (1)

$H_2 + 1/2O_2 = H_2O$ $\Delta H_2 = -285$ kJ (2)

$CH_3COOH + 2O_2 = 2CO_2 + 2H_2O$ $\Delta H_3 = -869$ kJ (3)

The equation corresponding to the formation of acetic acid from its elements is

$$2C + 2H_2 + O_2 = CH_3COOH \qquad \Delta H_f = ?$$

This equation may be obtained from the three thermochemical equations on multiplying equations (1) and (2) by two and adding, finally subtracting equation (3) :

$2 \times (1)$ $2C + 2O_2 = 2CO_2$

$2 \times (2)$ $2H_2 + O_2 = 2H_2O$

Adding $2C + 3O_2 + 2H_2 = 2CO_2 + 2H_2O$

Subtract (3) $CH_3COOH + 2O_2 = 2CO_2 + 2H_2O$

Giving, after re-arranging $2C + O_2 + 2H_2 = CH_3COOH$

$2 \times (1)$ $2 \times \Delta H_1 = -790$ kJ

$2 \times (2)$ $2 \times \Delta H_2 = -570$ kJ

Adding $\Delta H = -1360$ kJ

Subtract (3) $\Delta H_3 = -896$ kJ

Giving $\Delta H_f = -2360 - (-869) = -491$ kJ

Problem 24:

Heats of neutralization (ΔH) of NH_4OH and HF are – 51.5 and 68.6 kJ respectively. Calculate their heats of dissociation.

Solution:

We have

(i) $HCl(aq) + NaOH(aq) \rightarrow NaCl(aq) + H_2O$; $\Delta H = -57.3$ kJ

(ii) $HCl(aq) + NH_4OH(aq) \rightarrow NH_4Cl(aq) + H_2O$; $\Delta H = -51.5$ kJ
(Weak base)

∴ The heat of dissociation of NH4OH,

$$\Delta H = 51.5 - (57.3) = 5.8 \text{ kJ.}$$

Similarly we have

$HF(aq) + NaOH(aq) \rightarrow NaF + H2O;$

$\Delta H = -68.6$ kJ

∴ The heat of dissociation of HF,

$\Delta H = -68.6 - (57.3) = -11.3$ kJ.

Problem 25:

Calculate the heat of reaction at 25°C for the reaction :

$$C_2H_4(g) + H_2(g) \rightarrow C_3H_6(g)$$

given that the heats of combustion of ethylene, hydrogen and ethane at 25oC are 337.0, 68.4 and 373.0 kcal respectively.

Solution:

The thermochemical equations for the combustion of ethylene, hydrogen and ethane may be written as

(i) $C_2H_4 + 3O_2 \rightarrow 2CO_2 + 2H_2O \quad \Delta H = -337$ kcal

(ii) $H_2 + 1/2O_2 \rightarrow H_2O \quad \Delta H = -68.4$ kcal

(iii) $C_2H_6 + 31/2O_2 \rightarrow 3CO_2 + 3H_2O \quad \Delta H = -337$ kcal

$$(i) + (ii) - (iii) \Rightarrow C_2H_4 + H_2 - C_2H_6$$

Heat of reaction = – 337 – 68.4 + 373 = -- 32.3 kcal.

Problem 26:

For the hypothetical reaction

$$2A(g) \rightarrow A_2(g)$$

$\Delta C_p/J\ k^{-1}\ mol^{-1} = 1.00 + (2.00 \times 10\text{–}3/k)\ T$ and $\Delta H^o_{298}k = -5.00$ kJ mol^{-1}. Estimate the temperature at which $\Delta H^o = 0$ for this reaction at constant pressure.

Solution:

We have $\int_{\Delta H^o}^{0} d(\Delta H) = \int_{298\,k}^{T} (\Delta C_p)\, dT$

$$= \int_{298\,k}^{T} \left[1.00 + \left(\frac{2.00 \times 10^{-3}}{k}\right) T\right] J\ k^{-1}\ mol^{-1}\ dT$$

$0 - \Delta H^o = (1.00\ J\ k^{-1}\ mol^{-1})(T - 298\ k)$

$$+\left(\frac{2.00\times10^{-3}}{2k}\text{ J k}^{-1}\text{ mol}^{-1}\right)(T^2-298\text{ k}^2)$$

$$5000 = (1.00\text{ k}^{-1})(T - 298\text{ k}) + (10^{-3}\text{ k}^{-2})(T^2 - 298\text{k}^2) = 0$$

or $$10^{-3}\left(\frac{T}{K}\right)^2+\left(\frac{T}{K}\right)-298^2\times 10\text{–}3-298-5000=0$$

or $$10^{-3}\left(\frac{T}{K}\right)^2+\left(\frac{T}{K}\right)-5209=0$$

Solving for T/K, we get

$$\left(\frac{T}{K}\right)=\frac{-1+\left(1^2+4\times10^{-3}\times5209\right)^{1/2}}{2\times10^{-3}}$$

$$T=\frac{-1+4.67}{2\times10^{-3}}k=1836\text{ k}$$

Problem 27:

Compute the heat of reaction at 1000°C for

$1/2H_2(g) + 1/2\ Cl_2(g) \rightarrow HCL(g)\quad \Delta H^o_{298k} = 92.236\ k\ J\ mol^{-1}$

$C_p^o(H_2, g) = 29.0284 - 0.8355 \times 10^{-3}\ T + 2.0097 \times 10^{6}\ T^2.$

$C_p^o(Cl_2, g) = 31.6555 + 1.0134 \times 10^{-2}\ T - 4.0337 \times 10^{6}\ T^2.$

$C_p^o(HCl, g) = 28.1359 + 1.8074 \times 10^{2}\ T + 1.5453 \times 10^{-6}\ T^2.$

Solution:

$$\Delta H^o_{1273k} = \Delta H^o_{298k} + \int_{298k}^{1273k}(\Delta C_p)\,dT$$

where $\Delta C_p = \Delta C_p^o(HCl, g) - 1/2\ C_p^o(H_2, g) - 1/2C_p^o(Cl_2, g)$

$= (28.1359 - 1/2 \times 29.0284 - 1/2 \times 31.6555)$

$+ (1.8078 + 1/2 \times 0.8355 - 1/2 \times 10.134) \times 10^{-3}\ T$

$+ (1.5453 - 1/2 \times 2.0097 + 1/2 \times 4.0337) \times 10^{-6}\ T^2$

$= -2.2061 - 2.8415 \times 10^{-3}\ T + 2.5573 \times 10^{-6}\ T^2$

Hence,

$$\Delta H^{o}{}_{1273k} = \left[\begin{array}{l} -92.236\times10^{3}+(-2.2061)(1273-298)-\left(2.8415\times10^{-3}\right) \\ \left(\frac{1273^{2}}{2}-\frac{298^{3}}{2}\right)+\left(2.5573\times10^{-6}\right)\left(\frac{1273^{2}}{2}-\frac{298^{3}}{2}\right) \\ \end{array}\right]$$

J mol⁻¹

$= -\ 92236 - 2151.0 - 2176.2 + 1736.0$

$= -\ 94827 \text{ J mol}^{-1} = -\ 94.827 \text{ mol}^{-1}$

Problem 28:

The combustion of 1 g of benzene in a bomb calorimeter evolves 41.746 kJ of heat at 25°C, (i) What is ΔU° for combustion of benzene ? (ii) Calculate heat of formation of benzene.

Solution:

ΔH_f^{o} (CO_2, g) = – 393.129 kJ mol^{-1}

ΔH_f^{o} (H_2, l) = – 285.577 kJ mol^{-1}

The molar mass of benzene is 78 g mol^{-1}.

Hence,

$$\Delta U^{o} = -\ (41.746 \text{ kJ})\left(\frac{78 \text{ g mol}^{-1}}{1\text{g}}\right) = -\ 3256.16 \text{ kJ mol}^{-1}$$

The chemical equation of combustion is

$$C_6H_6(l) + \frac{15}{2}O_2(g) \rightarrow 6CO_2(g) + 3H_2O(l)$$

$$\Delta V_g = 6 - \frac{15}{2} = -\frac{3}{2}$$

Hence, $\Delta H^{o} = \Delta U^{o} + (\Delta V_g)RT = [(-\ 3256.16 \times 10^3)$

$+ (-\ 3/2)(8.314 \times 298)]$ J mol^{-1}

$= -\ 3259.9 \times 10^3$ J mol^{-1}.

Finally, we have

$$\Delta H^{o} = 6\Delta H_f^{o} + (CO_2, g) + 3\Delta H_f^{o}(H_2O, l) - \Delta H_f^{o}(C_6H_6, l)$$

which gives

$\Delta H_f^o (C_6H_6, l) = 6\Delta H_f^o +(CO_2, g) + 3\Delta H_f^o (H_2O,l) - \Delta H_f^o$

$= [-6 \times 393.129 - 3 \times 285.577 + 3259.9]$ kJ mol^{-1}

$= -44.35$ mol^{-1}

Problem 29:

From the following data, calculate ΔH_f at 298.15 k for (i) $OH^-(aq_)$ and (ii) $Na^+(aq)$:

ΔH_f^o values is kJ mol^{-1} at 298.15 k for

$H_2O,(l) = -285.8$ $HCl (aq) = -167.4$

$NaOH(aq) = -469.6$ $NaCl(aq) = -407.0$

Solution:

(i) $H_2(g) + 1/2\ O_2(g) \rightarrow H_2O(l)$

$\Delta H^o = -285.8$ kJ mol^{-1}

(ii) $1/2H_2(g)+1/2\ Cl_2(g)+aq \rightarrow H^+(aq)+Cl^-(aq)$

$\Delta H^o = -167.4$ kJ mol^{-1}

(iii) $Na(s) + 1/2\ O_2(g) + 1/2\ H_2(g) + aq \rightarrow Na^+(aq) + OH^-(aq)$

$\Delta H^o = -469.6$ kJ mol^{-1}

(iv) $Na(s) + 1/2\ Cl_2(g) + aq \rightarrow Na^+(aq) + Cl^-(aq)$

$\Delta H^o = -407.0$ kJ mol^{-1}

Carrying out the manipulation

Eq. (iii) – Eq. (iv) + Eq. (ii)

We get

(v) $1/2O_2(g) + H_2(g) + aq \rightarrow H^+(aq) + OH^-(aq)$

Hence, its enthalpy change becomes

$\Delta H^o = \Delta H^o_{(m)} - \Delta H^o_{(iv)} + \Delta H^o_{(ii)}$

$= (-469.6 + 407.0 - 167.4)$ kJ mol^{-1}

$= -230.0$ kJ mol^{-1}

The enthalpy change of Eq. (v) is also given as

$\Delta H = \Delta H_f^o(H^+) + \Delta H_f^o(OH^-) - 1/2\ \Delta H_f^o (O_2) - \Delta H_f^o (H_2)$

$= 0 + \Delta H_f^o(OH^-) - 0 - 0 = \Delta H_f^o (OH^-)$

Thus ΔH_f^o (OH^-, aq) = – 230.0 kJ mol^{-1}

From Eq, (iii), we get

$\Delta H = \Delta H_f^o(Na^+) + \Delta H_f^o(OH^-) - \Delta H_f^o(Na) - 1/2\,\Delta H_f^o(O_2) - 1/2\,\Delta H_f^o(H_2)$

– 469.6 kJ mol^{-1} = $\Delta H_f^o(Na^+)$ = – 230.0 kJ mol^{-1}

which gives

$\Delta H_f^o(Na^+)$ = (– 469.6 + 230.0) kJ mol^{-1}

= – 239.6 kJ mol^{-1}.

Problem 30:

Calculate ΔH^o *for the reaction at 298 K*

$Ba^{2}\ (aq) + SO_4^{2} \rightarrow BaSO_4(s)$

Given that

$\Delta H_f^o\ [Ba^{2-}\ (aq)]$ = – 538.36 kJ

$\Delta H_f^o\ [SO_4^{2-}\ (aq)]$ = – 907.51 kJ

$\Delta H_f^o\ [BaSO_4\ (s)]$ = – 1465.24 kJ

Solution:

For the reaction,

$Ba^{2+}\ (aq) + SO_4^{2-} \rightarrow BaSO_4(s)$

We have $\Delta H^o = \Delta H_f^o\ [BaSO_4(s)] - \Delta H_f^o\ [Ba^{2+}\ (aq)] - \Delta H_f^o\ [SO_4^{2-}\ (aq)]$

On substituting the values, we get

ΔH^o = – 1465.24 kJ – (– 538.36 kJ – 907.51 kJ) = – 19.37 kJ

Problem 31:

Calculate the bond enthalpy for a C – O bond in methanol from the following data :

(i) $C(s) + 2H_2(g) + 1/2O_2 \rightarrow CH_3OH(g)$; ΔH^o = – 200 kJ mol^{-1}

(ii) $C(s) \rightarrow C(g)$; ΔH^o = – 716.8 kJ mol^{-1}

(iii) $2H_2(g) \rightarrow 4H(g)$; ΔH^o = –872.0 kJ mol^{-1}

(iv) 1/2O2(g) → O(g); ΔH^o = –249.0 kJ mol^{-1}

The bond enthalpy for C – H bond is 403 kJ mol^{-1} *and for O–H bond it is 463.6 kJ* mol^{-1}.

Solution:

The enthalpy change for the dissociation of CH_3OH is as follows :

$CH_3OH(g) \rightarrow C(g) + 4H(g) + O(g)$; $\Delta H^o = ?$

On adding (ii), (iii) and (iv), we get

$C(s) + 2H_2(g) + 1/2O_2(g) \rightarrow C(g) + 4H(g) + O(g)$,

$\Delta H^o = 1837.8$ kJ

On subtracting equation (i) from the above equation, we get

$CH_3OH(g) \rightarrow C(g) + 4H(g) + O(g)$;

$\Delta H^o = 2037.8 kJ\ mol^{-1}$

As methanol has three C – H, one C – O bond and one O – H bond, it means that

$$2037.8 \text{ kJ} = 3\ (\Delta H^o_{C-H}) + (\Delta H^o_{OH}) + 1\ (\Delta H^o_{C-O})$$

$$= 3\ (413) + (463.6) + ((\Delta H^o_{C-O})$$

or $$(\Delta H^o_{C-O}) = 335.2 \text{ kJ mol}^{-1}$$

Problem 32:

The enthalpy of neutralisation of ammonium hydroxide by hydrochloric acid is 51.46 kJ mol^{-1}. Calculate the enthalpy of ionisation of ammonium hydroxide.

Solution:

(i) $NH_2OH(aq)+H^+(aq) \rightarrow NH_4^+(aq)+H_2O(l)$; $\Delta H^o = -51.46 kJ\ mol^{-1}$

(ii) $H^+(aq) + OH^-(aq) \rightarrow H_2O(l)$; $\Delta H^o = -57.32 kJ\ mol^{-1}$

(iii) $NH_2OH(aq) \rightarrow NH_4^+(aq) + OH-(aq)$; $\Delta H^o = \Delta H^o_{Ionisation}$

On adding (ii) and (iii), we get

$$NH_2OH(aq) + H^+(aq) \rightarrow NH_4^+(aq) + H_2O(l)$$

or $$\Delta H^o = -\ 57.32 + \Delta H^o_{Ionisation}$$

From Eq. (i), we have

$$\Delta H^o = -\ 51.46 \text{ kJ mol}^{-1}$$

Hence $(-\ 57.32 + \Delta H^o_{Ionisation})$ kJ mol^{-1} = – 51.46 kJ mol^{-1}

$\Delta H^o_{Ionisation} = +\ 5.86$ kJ mol^{-1}.

Problem 33:

If E_{C-C} is 34 kJ mole^{-1} and E_{C-H} is 415 kJ mole^{-1}, calculate the heat of formation of propane. The heats of atomization of carbon and hydrogen are 716 kJ mole^{-1} and 433 kJ mole^{-1} respectively.

Solution:

The heat of formation is the sum of the heats of atomization and bond energies. For propane, the heats of atomization as

$$3C_s = 3C_g; \qquad \Delta H = 3 \times 716 = 2148 \text{ kJ}$$

$$4H_{2,g} = 8H_g; \qquad \Delta H = 4 \times 433 = 1732 \text{ kJ}$$

The bond energies are

$$2E_{C-C} = 2 \times -344 = -688 \text{ kJ}$$

$$2E_{C-H} = 8 \times -415 = -3320 \text{ kJ}$$

Adding,

$$3C + 4H_2 = C_3H_8; \qquad \Delta H_f = 2148 + 1732 - 688 - 3320$$

$$= -128 \text{ kJ mole}^{-1}$$

Problem 34:

The heats of formation of PCl_3 and PH_3 are c 306 kJ mole^{-1} and + 8 kJ mole^{-1} respectively, and the heats of atomization of phosphorus, chlorine and hydrogen are given by :

$$P_s = P_g \qquad \Delta H = 314 \text{ kJ mole}^{-1}$$

$$Cl_{2.g} = 2Cl_g \qquad \Delta H = 242 \text{ kJ mole}^{-1}$$

$$H_{2.g} = 2H_g \qquad \Delta H = 433 \text{ kJ mole}^{-1}$$

Calculate E_{P-Cl} and E_{P-H}

Solution:

For phosphorus trichloride, thermochemical equations are :

$$P_s = P_g \qquad \Delta H_1 = 314 \text{ kJ}$$

$$3/2\ Cl_{2.g} = 3Cl_g \qquad \Delta H_2 = 363 \text{ kJ}$$

$$P_g + 3Cl_g = P_{Cl3} = 2H_g \qquad \Delta H_3 = ?$$

Also,

$$P_s + 3/2\ Cl_{2.g} = PCl_3 \qquad \Delta H_f = -306 \text{ kJ}$$

Equating the two routes leading to the formation of PCl_3,

$$\Delta H_f = \Delta H_1 + \Delta H_2 + \Delta H_3$$

$$-306 = 314 + 363 + \Delta H_3$$

$$\Delta H_3 = -983 \text{ kJ.}$$

This is energy released on the formation of three P – Cl bonds, hence,

$$E_{P-Cl} = \frac{983}{3} = 328 \text{ kJ mole}^{-1}$$

$$P_s = P_g \qquad \Delta H_1 = 314 \text{ kJ}$$

$$3/2\ H_{2,g} = 3H_g \qquad \Delta H_2 = 649 \text{ kJ}$$

$$P_2 + 3H_g = PH_3 \qquad \Delta H_3 = ?$$

Also,

$$P_s + 3/2\ H_{2,g} = PH_3 \qquad \Delta H_f = 8 \text{ kJ}$$

From which

$$\Delta H_3 = -958 \text{ kJ}$$

$$E_{P-H} = \frac{958}{3} = 319.33 \text{ kJ}$$

Problem 35:

100g of anhydrous $CuSO_4$, when dissolved in excess of water produced 42 kJ of heat. The same amount of $CuSO_4$. $5H_2O$ on dissolving in large excess of water absorbed 4.60 kJ. What is the heat of hydration $CuSO_4$?

Solution:

$$100\text{g of } CuSO_4 = \frac{100}{159.5} \text{ mole}$$

$$\text{Heat of solution of } CuSO_4 \text{ per mole} = \frac{42 \times 159.5}{100} = -66.78 \text{ kJ}$$

$$100 \text{ g of } CuSO_4.\ 5H_2O = \frac{100}{249.5} \text{ mole}$$

$$\text{Heat of solution of } CuSO_4.\ 5H_2O \text{ per mole} = 4.6 \times \frac{249.5}{100} = 11.48 \text{ kJ}$$

Thus,

(i) $CuSO_4.\ 5H_2O + aq \rightarrow CuSO_4(aq) + 66.78$ kJ

(ii) $CuSO_4.\ 5H_2O + aq \rightarrow CuSO_4(aq) - 11.48$ kJ

On subtracting (i) from (ii), we get

$$CuSO_4 + 5H_2O \rightarrow CuSO_4\ 5H_2O + 78.26 \text{ kJ}$$

∵ Enthalpy of hydration of $CuSO_4$ $\Delta H = -78.26$ kJ.

Problem 36:

Heat of combustion of glucose at constant pressure at 17oC was found to be –651,000 cal. Calculate the heat of combustion of glucose at constant volume considering water to be in the gaseous state.

Solution:

$$\underset{\text{negligible volume}}{C_6H_{12}O_6(s)} + \underset{6 \text{ mole}}{6O_2(g)} \rightarrow \underset{6 \text{ mole}}{6CO_2(g)} + \underset{6 \text{ mole}}{H_2O(g)}$$

$$\Delta n = 12 - 6 = 6$$

$$\Delta H = \Delta E + \Delta nRT$$

$$\Delta E = \Delta H - \Delta nRT = -651000 - 6 \times 2 \times (273 + 17)$$

$$= 651000 - 3480 = -654480 \text{ cal}$$

Hence heat of combustion of glucose at constant volume and at 17°C is – 654,480 cal.

Problem 37:

The combustion of methane takes place according to the following equation :

$$CH_4(g) + 2O_2(g) \rightarrow CO_2(g) + 2H_2O(l)\ ; \qquad \Delta H = -890.3 \text{ kJ}$$

(a) How many grams of methane would be required to produce 89.03 kJ of heat of combustion ?

(b) How many grams of carbon dioxide would be formed when 89.03 kJ of heat is evolved ?

Solution:

(a) Molecular weight of $CH_4 = 1 \times 12 + 4 \times 1$

$$= 12 + 4 = 16 \qquad [\because C = 12, H = 1]$$

∵ 890.3 kJ of energy is liberated by the combustion of 1 mole of 16 g of methane.

∴ 890.3 kJ of energy is liberated by the combustion of

$$16 \times \frac{89.03}{890.3} = 1.6 \text{ g of methane.}$$

= 1.6 g of methane.

(b) Molecular weight of CO_2 = 1 × 12 + 2 × 16

= 12 + 32 = 44

[∵ C = 12, O = 16]

∵ 890.3 kJ of energy is liberated to produce 1 mole or 44 g of CO_2

∴ 890.3 kJ of energy is liberated to produce

$$44 \times \frac{89.03}{890.3}$$

or 4.4 g of CO_2

= 4.4g of carbon dioxide.

Problem 38:

The formation of SO_3 from SO_2 and oxygen,

SO_2 (g) + 1/2 O_2 (g) → SO_3 (g)

is exothermic by 97030 *at* 298 *K and 1 atm measured in a bomb calorimeter. What is the value of* Δ H?

Solution:

The above equation can be written as,

SO_2 (g) + 1/2 O_2 (g) → SO_3 (g) ΔE (298 K) = – 97030 J.

We know

ΔH = ΔE + ΔvRT

Here $\Delta v = 1 - \left(1+\frac{1}{2}\right) = -\frac{1}{2}$

∴ ΔH = – 97030 – 1/2 (8.314) (298) = – **98270 J.**

Problem 39:

Find ΔE° *for the following reactions at 1 atm and* 298 K

C (s) + O_2(g) → CO_2(g) ΔH° (298 K) = – 393.5 kJ mol^{-1}

C (s) + 1/2 O_2 (g) → CO (g) ΔH° (298 K) = – 110.5 kJ mol^{-1}

CH_4(g) + 2O_2(g) → CO_2(g) + 2H_2O(l) ΔH° (298 K)

= – 890.35 kJ mol^{-1}

$AgNO_3$ (s) + HCl (g) → AgCl (s) + HNO_3 (l)

ΔH° (298 K) = – 85.3 kJ mol^{-1}.

Solution:

Since R = 8.314 $JK^{-1}mol^{-1}$

and T = 298 K in the present case

RT = (8.314 JK^{-1} mol^{-1}) × (298K)

= 2477.57 J mol^{-1} = 2.477 kJ mo^{-1}.

The moles of only gaseous species are to be considered

$\Delta H^\circ = \Delta E^\circ + \Delta v$ RT gives $\Delta E^\circ = \Delta H^\circ - \Delta v$ RT

(i) C(s) + O_2(g) → CO_2 (g) ΔH°(298K) = – 393.5 kJ mol^{-1}

$\Delta v = 1 - 1 = 0$

∴ ΔH° (298 K) = ΔE° (298 K)

= 393.5 KJ mol^{-1}

(ii) C (s) + 1/2 O_2 (g) → CO (g) ΔH° (298 K)=– 110.5kJ mol^{-1}

$$\Delta v = 1 - \frac{1}{2} = \frac{1}{2}$$

$$\therefore \Delta E^\circ \text{ (298 K)} = -110.5 \text{ kJ mol}^{-1} - \left(\frac{1}{2}\right) \times (2.477 \text{ kJ mol}^{-1})$$

= **–111.738 kJ mol^{-1}**

(iii) CH_4 (g) $2O_2$(g) → CO_2 (g) + $2H_2O$(l) ΔH° (298 K)

= – 890.35 kJ mol^{-1}

$\Delta V = 1 - (1 + 2) = -2$

ΔE° (298 K) = (– 890.35 kJ mol^{-1}) + 2 × (2. 4 77kj mol^{-1})

= – **885.39 kJ mol^{-1}**

(iv) $AgNO_3$ (s) + HCl (g) → AgCl (s) + HNO_3(l)

ΔH° (298 K) = – 85.3 kJ

$\Delta v = 0 - 1 = -1$

ΔE° (298 K) = – (85.3 kJ mol^{-1}) – (–1) × (2.477 kJ mol^{-1})

= – **82.823 kJ mol^{-1}**.

Problem 40:

Compute the heat change in the following reaction

$COCl_2 + 2H_2S \rightarrow 2HCl + H_2O + CS_2$ *given that*

(i) $COCl_2 + H_2S \rightarrow 2HCl + COS$ ΔH° (298) = – 78.705 kJ

(ii) $COS + H_2S \rightarrow H_2O + CS_2$ ΔH° (298) = – 78.705 kJ

Solution:

On adding equation (i) and (ii) and cancelling the identical terms we get the desired result as

$COCl_2 + H_2S \rightarrow 2HCl + COS \ \Delta H^\circ_{(i)} = -78.705$ kj

$$\frac{COS + H_2S \rightarrow H_2O + CS_2}{COCl_2 + 2H_2S \rightarrow 2HCl + H_2O + CS_2} \quad \frac{\Delta H^\circ_{(ii)} = +3.420\text{kj}}{\Delta H^\circ(298) = \mathbf{-75.285kJ}}$$

Problem 41:

At 25°C the enthalpy changes for the dissolution of the two forms of glucose and their mutarotation are respectively given by

(i) α-D-*glucose* (s) → α-D-*glucose* (aq) ΔH(i) = 10.72 *kJ*

(ii) β-D-*glucose* (s) → β-D-*glucose* (aq) ΔH(ii) = 4.68 *kJ*

(iii) α-D-*glucose* (aq) → β-D-*glucose* (aq) ΔH(iii) = 1.16 *kJ*

Calculate ΔH *for* α-D-*glucose* (s) → β-D-*glucose* (s) ΔH = ?

Solution:

Reverse equation (ii) and then add this and eqn. (i) and, (iii) cancel the identical terms :

(iv) β-D-*glucose* (aq) → β-D-*glucose* (s) ΔH(iv) = – 4.68 *kJ*

(v) α-D-*glucose* (aq) → β-D-*glucose* (aq) ΔH(iii) = 1.16 *kJ*

$$\text{(i)} \ \frac{\alpha-\text{D}-\text{glucose (s)} \rightarrow \alpha\text{-D-glucose (aq)}}{\alpha\text{-D-glucose (s)} \rightarrow \beta\text{-D glucose (s)}} \quad \frac{\Delta H(i) = 10.72\text{kJ}}{\Delta H = \mathbf{4.88kJ}} \ \text{Add}$$

Problem 42:

Compute the standard heat of formation of methane using the following data:

(i) $CH_4(g) + 2O_2(g) \rightarrow CO_2(g) + 2H_2O(l)$ ΔH° (298K)=– 890.35 kJ

(ii) $H_2(g) + 1/2\ O_2(g) \rightarrow H_2O(l)$ ΔH° (298K) = – 285.84 kJ

(iii) C (*graphite*) + O_2 (g) → CO_2 (g) ΔH° (298K) = – 393.51 kJ

Solution:

Our aim is to get DH° for the following reaction

C (graphite) + $2H_2$ (g) → CH_4(g) ΔH° (298) = ?

Reverse equation (i) and multiply equation (ii) by two.

(iv) CO_2 (g) + $2H_2O$ (l) → CH_4 (g) ΔH° (298K) = – 285.84 kJ

(v) $2H_2$ (g) + O_2(g) → $2H_2O$(l) ΔH° (298K) = – 571.68 kJ

Add equations (iii), (iv) and cancel the common terms on both sides of the equations. This gives

C (graphite) + $2H_2$ (g) → CH_4 (g) ΔH° (298K) = – **74.84 kJ**

Thus the enthalpy change for the formation of methane from its elements is calculated to be 74.84 kJ.

Problem 43:

Calculate the enthalpy change for the following reaction at 25°C.

$$C_2H_6\ (g) + \frac{7}{2}O_2\ (g) \rightarrow 2CO_2(g) + 3H_2O(l)$$

Use the enthalpy of formation of the reactants and products from literature.

Solution:

Enthalpy change for any reaction is given as

$\Delta H^\circ = \Sigma H^\circ_f\ (\text{products}) - \Sigma\Delta H^\circ_f\ (\text{reactants})$

For the given reaction

$\Delta H^\circ = 2\Delta H^\circ_f\ (CO_2) + 3\Delta H^\circ_f\ (H_2O) - (\Delta H^\circ_f\ C_2H_6 + 7/2\ \Delta H^\circ_f\ O_2)$

$$= -2 \times 393.5 - 3 \times 285.8 - \left(-84.5 + \frac{7}{2} \times 0\right)$$

$= -787.0 - 857.4 + 84.5$

$= -$ **1559.9 kJ.**

Problem 44:

Calculate the heat of formation of benzene from the following data:

$$C_6H_6(l) + \frac{15}{2}\ O_2(g) \rightarrow 6CO_2(g) + 3H_2O(l)\quad \Delta H^\circ = -3303\ kJ$$

$\Delta H^o_f (CO_2) = - 393.5$ kJ; $\Delta H^o_f (H_2O) = - 285.8$ kJ.

Solution:

For the given reaction

$$\Delta H^o = 6 \times \Delta H^o_f (CO_2) + 3\ \Delta H^o_f (H_2O) - \Delta H^o_f (C_6H_6) - \frac{15}{2} \Delta H^o_f (O_2).$$

$$\Delta H^o_f\ C_6H_6 = 6 \times \Delta H^o_f (CO_2) + 3 \times \Delta H^o_f (H_2O) - \Delta H^o$$

$$= - 6 \times 393.5 - 3 \times 285.8 + 3303 = \mathbf{84.6\ kJ.}$$

Problem 45:

The heat of solution DH at different molalities m of an aqueous solution of a solute X are given below.

m	1.00	1.80	2.82	4.25	6.37	9.95
ΔH(298K)/kJ	–34.2	–56.7	–88.3	–126.0	–174.0	–229.0

Plot ΔH against the molality m, and determine the differential heat of solution of the solute X in a 3-molal solution.

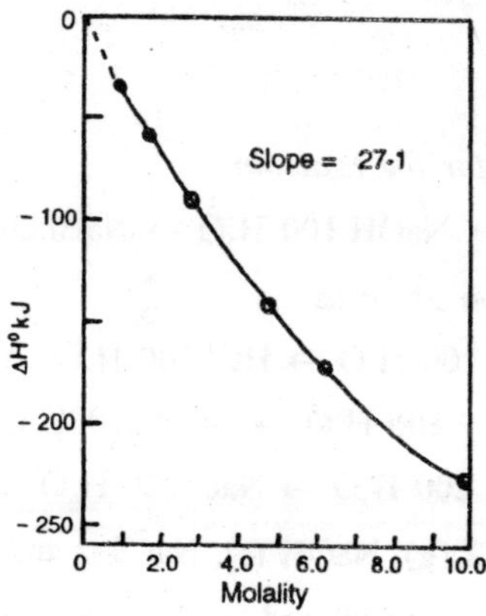

Solution:

ΔH values are plotted against molality. Now draw a tangent on the curve at a point when molality is 3. Extend this tangent to meet the Δ H axis, and find out the slope of the tangent. It is –27.1kJ molal^{-1}. Since the variation of ΔH with molality (m_2) is given by

$$\Delta H = n_1\ \Delta H_{1\cdot\ m} + n_2\ \Delta H_{2},\ m = 55.55\ \Delta H_{1m} + m_2\ \Delta H_{2,\ m}$$

The slope of the tangent is equal to $\Delta H_{2', m}$ *i.e.*, $\Delta H_{2', m}$ = **–27.1kJ** molal^{-1}.

Problem 46:

Calculate the integral heat of solution of one mole HCl(g) in 100 H2O (l), using the information that

$\Delta H^o_{f1(HCl)}$ = – 92.3kJ

$\Delta H^o_{f2(HCl\ in\ 100\ H2O)}$ = – 168.0 kJ

Solution:

The dissolution process can be represented by a chemical equation as

HCl (g) + 100 H_2O → HCl.100 H_2O

$\Delta H^o = \Sigma\Delta H^o_f$ (products) – $\Sigma\Delta H^o_f$ (reactants)

Since, 100 H_2O is on both sides of the above equation, we will not consider the enthalpy of formation of water. Therefore,

ΔH^o = – 168.0 – (–92.3)

= **– 75.7 kJ.**

Problem 47:

Calculate ΔH for the reaction

HCl.100 H_2O + NaOH.100 H_2O → NaCl.200 H_2O + H_2O(l)

The information given as

(i) HCl (g) + 100 H_2O → HCl.100 H_2O ΔH (i) = –75.7 kJ

(ii) NaOH (s) + 100 H_2O → NaOH.100H_2O ΔH (ii) = –40.9 kJ

(iii) Nacl (s) + 200 H_2O → Nacl.200 H_2O ΔH (iii) = – 4.26 kJ

ΔH_f^o's for HCl (g), NaOH (s), NaCl(s) and H_2O (l) are

–92.3, – 426.7, – 411.0 and – 285.8 kJ *respectively.*

Solution:

The given reaction is

(a) HCl.100 H_2O + NaOH.100 H_2O → NaCl.200 H_2O + H_2O(l)

$\Delta H = \Sigma\Delta H^o_f$ (products) – $\Sigma\Delta H^o_f$ (reactants)

ΔH for this reaction can be calculated if the ΔH^o_f data for HCl in 100 H_2O; NaOH in 100 H_2O, NaCl in 200 H_2O and H_2O (l) are known.

From equation (i), we get

ΔH°_f (HCl.100 H_2O) = –75.7 –92.3 = – 168.0 kJ

Equation (iii) gives

ΔH°_f (NaOH.100 H_2O) = ΔH(ii) + ΔH°_f [NaOH(s)]

= – 40.9 – 426.6 = – 467.5 kJ

From equation (iii), we have

ΔH°_f (NaCl.200H_2O) = + 4.26 – 411.0 – 406.74 kJ

Also ΔH°_f H_2O = – 285.8 kJ

Using these data we have

ΔH(a) = (– 406.74 –285.8) – (– 168.0 – 467.5)

= 692.54 + 635.50 = **–57.04 kJ.**

Heat of Neutralization

It is change in enthalpy when one mole H^+ ions are completely neutralized by a base in very dilute solutions. The dilute solution is necessary so that on mixing the acid and base the thermal change due to dilution is negligible. The heat of neutralization of hydrochloric acid and sodium hydroxide is –57.0 kJ mol–1 at 25°C and 1 atm pressure. In the case of weak acid and strong base the heat of neutralization is less than 57.0 kJ. The difference in heat of neutralization values of strong acid – strong base, and weak acid – strong base is attributed to the heat of ionization of weak acid.

Heat of neutralization of strong and strong base. Consider the neutralization of aqueous solution of HCl by NaOH. The reaction may be represented by

H^+ (aq) + Cl^- (aq) + Na^+ (aq) + OH^- (aq)

→ Na^+ (aq) + Cl^- (aq) + H_2O (l)

Since Na^+ (aq) and Cl^- (aq) are common on both sides, the neutralization of strong acid and strong base may be represented by

H+ (aq) + OH^- (aq) → H_2O(1)

Thus heat of nutralization of strong acid and strong base (mono-) is simply the heat of formation of H_2O from H^+ and OH^- ions which is fixed at a given temperature and pressure. This leads to the conclusion that the heat of neutralization of strong monobasic acid with strong base is the same. Heat of neutralization of a given acid-base pair can experimentally be found by using a calorimeter. The heat of ionization of a weak acid of a base can also be measured by calorimetric method. For example, the heat of ionization of a weak acid (acetic acid) will be equal to the heat of reaction between sodium acetate and hydrochloric acid at the a given temperature. This quantity will also be equal to the difference between the heat of neutralization of NaOH – HCl pair and acetic acid – NaOH pair. This is shown as follows:

(i) CH_3COOH *** $CH_3\ COO^- + H^+$ ΔH (ionization) = ?

(ii) $NaOH + HCl \rightarrow Na^+\ Cl^- + H_2O$ $\Delta H_1 = xkJ$

(iii) $NaOH + CH_3COOH \rightarrow CH_3COONa + H_2O$ $\Delta H_2 = ykJ$

Reverse equation (ii) to get (iv)

(iv) $Na^+\ Cl^- + H_2O \rightarrow NaOH + HCl$ $\Delta H_3 = -\ xkJ$

Add (iii) and (iv)

$CH_3COOH + Na^+ + Cl^- \rightarrow CH_3COONa + HCl$

$\rightarrow CH_3\ COO^- + Na^+ + H^+ + Cl^-$

or $CH_3\ COOH$ ***$CH_3COO^- + H^+$ $\Delta H(ion) = \Delta H_2 + \Delta H_3$

$= (y - x)$ kJ

Problem 48:

The heat of neutralization of a strong acid with a strong base is – 57.36 kJ mol⁻¹. If the heat of neutralization of a weak acid HA with a strong base be –42.02 kJ mol⁻¹ compute the heat of ionization of the weak acid.

Solution:

The neutralization of strong acid with strong base may essentially be represented as

$H^+\ (aq) + OH^-\ (aq) \rightarrow H_2O(l)$ $\Delta H = -\ 57.36\ kJ\ mol^{-1}$

The neutralization of a weak acid with a strong base may be considered to be the resultant of two successive steps.

$HA\ (aq) \rightarrow H+\ (aq) + A^-\ (aq)$ $\Delta H = \Delta H$ (ionization)

$H + (aq) + OH^-(aq) \rightarrow H_2O(l)$ $\Delta H = -\ 57.36$ kJ

$HA(aq) + OH^-(aq) \rightarrow H_2O(l) + A^-(aq)$

$\Delta H = 42.02$ kJ

So that the enthalpy change is

ΔH (neutralization) = ΔH (ionization) – 57.36 kJ mol^{-1}

and ΔH (ionization) = – 42.02 + 57.36 = **15.34** kJ mol^{-1}.

Heat of Formation of Ions in Solution

The complete thermal data on aqueous solutions of electrolytes can be presented by tabulating heats of formation of ions at unit activity, which can be used to calculate the thermal changes for the reaction in which the ions participate.

Since in aqueous solutions the H^+ and OH^- ions are involved in one way or the other, first we shall consider the computation of the standard heats of formation of these ions.

The heat of formation of one mole of water from hydrogen and hydroxyl ions is essentially equal to the heat of neutralization of strong acid and strong base (–57.0 kJ mol^{-1}) and the standard heat of formation of water from its elements H2 and O_2 is –285.8 kJ mol^{-1}. That is

H^+ (aq) + OH^- (aq) $\rightarrow$ $H_2O(l)$ $\Delta H°$ (298K) = –57.0 kJ

or (i) H_2O (l) $\rightarrow$ H^+ (aq) + OH^- (aq) $\Delta H°$ (298K) = –57.0 kJ

and (ii) $H_2(g) + \frac{1}{2}O_2(g) \rightarrow H_2O(l)$ $\Delta H°$ (298K) = –285.8 kJ

Adding these equations we get

(iii) H_2 (g) + 1/2 O_2(g) $\rightarrow$ H^+ (aq) + OH^- (aq) $\Delta H°$ (298K)=– 228.8kJ

The sum of the heat of formation of H^+ and OH^- ions is – 228.8 kJ mol^{-1}. By convention the standard heat of formation of hydrogen ions in aqueous solution is taken as zero.

1/2 H_2 (g) ¬ H^+ (aq) + e^- $\Delta H°$ (298 K) = 0

With this convention equation (iii) directly gives the heat of formation of hydroxyl ions,

$\frac{1}{2}H_2(g) + \frac{1}{2}O_2(g)\ e^-$

$\rightarrow$ OH^-(aq) $\Delta H°$ (298K) = -- 228.8 kJ

The standard heat of formation of other ions is solution can be calculated in a similar manner.

Problem 49:

The standard heat of formation of HCl in water at 25°C is –168.0kJ. Calculate the standard heat of formation of chloride ions.

Solution:

(i) $\frac{1}{2}H_2\ (g) + Cl_2(g) + aq \rightarrow H^+\ (aq) + Cl^-\ (aq)$

$\Delta H^o\ (298K) = -168.0\ kJ$

(ii) $\frac{1}{2}H_2\ (g) \rightarrow H^+\ (aq) + e^-\ \Delta H^o\ (298\ K) = 0$

Subtracting (ii) from (i) we get

$\frac{1}{2}Cl_2\ (g) + aq + e^- \rightarrow Cl^-\ (aq)\ \Delta H^o\ (298\ K) = \mathbf{-168.0kJ.}$

Problem 50:

Calculate the standard heat of formation of sodium ions in aqueous solution using the standard heat of formation of NaOH in aqueous medium (ΔH^o_f = – 470.7kJ); $\Delta H^o_f\ (OH^-)$ = – 228.8 kJ.

Solution:

Given data :

(i) $Na(s) + \frac{1}{2}\ H_2\ (g) + \frac{1}{2}O_2\ (g) + aq \rightarrow Na^+\ (aq)\ + OH^-\ (aq)$

$\Delta H^o = -\ 470.7\ kJ$

(ii) $\frac{1}{2}\ H_2(g) + \frac{1}{2}O_2(g) + e^- \rightarrow OH^-(aq)\ \Delta H^o = -228.8kJ$

Subtracting equation (ii) from (i) we get

$Na(s) + (aq) + e^- \rightarrow Na^+\ (aq) + e^-\ \Delta H^o = \mathbf{-241.9kJ}$

Problem 51:

Examine the reactions tabulated below with the accompanying change in enthalpy :

(a) *Prepare Born-Haber cycles for the separate LiCl and NaCl and calculate their respective lattice energies.*

(b) *Using the following heats of solutions of LiCl and NaCl in water along with the lattice energy data obtained in part* (a), *calculate the heat of hydration of* $Li^{\cdot}$ *and* Na^{-} *ions separately and interpret the results*:

Process	*Reaction*	*Enthalpy changes (kJ)*
Sublimation :	(i) Li (s) → Li (g)	$\Delta H_{sub} = +161$
	(ii) Na (s) → Na (g)	$\Delta H_{sub} = +109$
Dissociation :	1/2 Cl (g) → Cl (g)	$1/2\Delta H_d = +122$
Electron-affinity :	Cl (g) + e^- → Cl^- (g)	$\Delta H_{aff} = -350$
Ionization :	(i) Li (g) → Li^+ + e^-	$\chi H_{IP} = +520$
	(ii) Na (g) → Na^+ + e^-	$\Delta H_{IP} = +496$
Compound	(i) Li (s) + $\frac{1}{2}Cl_2$ (g) → LiCl (s)	$\Delta H^o_f = -410$
formation :	(ii) Na (s) + $\frac{1}{2}Cl_2$(g) → NaCl(s)	$\Delta H^o_f = -411$

(i) LiCl (s) + aq → Li^+ (aq) + Cl– (aq) + 35.1 kJ

(ii) NaCl (s) + aq → Na^+ (aq) + Cl^- 4.3 kJ

Solution:

(a) Born-Haber cyclic is prepared as discussed for general case MX.

For lithium chloride: LiCl (s) → Li^+ (g) + Cl^- (g)

$\Delta H_C = 161 + 122 + 520 + 410$

= **863 kJ**

For sodium chloride : NaCl (s) → Na^+ (g) + Cl^+ (g)

$\Delta H_C = 109 + 122 + 496 - 350 + 411$

= 788 kJ

(b) Heat of hydration = Heat of solution – Crystal energy

Solution process : *For lithium chloride*

LiCl (s) + aq → Li^+ (aq) + Cl^- (aq) $\Delta H = -35.1$ kJ

Crystal formation process:

Li^+ (g) + Cl^- (g) → LiCl(s) $\Delta H = -35.1$ kJ

Adding the two equations, we get

$Li^+ (g) + Cl^- (g) \xrightarrow{aq} Li (aq) + Cl^- (aq)\ \Delta H_{hydration} = -898.1$ kJ

Similarly for NaCl

$Na^+ (g) + Cl^- (g) \xrightarrow{aq} Na+ (aq) + Cl^- (aq)\ \Delta H_{hydration} = -$ **783.7kJ**

The interaction of gaseous ions with water is called *hydration* and accompanying enthalpy change is called the hydration energy. The hydration energy for lithium ion (radius = 0.060 nm) is larger compared with the hydration energy of Na^+ (radius = 0.095 nm). This suggests that the interaction between Li^+ ion and water dipole is greater (ion– dipole interaction) than the ion – dipole interaction possibility between Na^+ ions and water molecules. This is obvious because there is a size difference between Na^+ ions and Li^+ ions so that the water dipoles are father away from the centre of the sodium ions but nearer to the Li^+ ions, whereas the net ionic charges are the same in both the cases. Therefore water molecules are held less strongly around the sodium ions than around the lithium ions.

The difference in the hydration energies of Li^+ ions and Na^+ ions is about 114 kJ mol^{-1}. However, the crystal energies differ only by about 75 mol^{-1}. The fact that lithium chloride dissolves with the evolution of heat (–35.1kJ) and sodium chloride dissolves with absorption of heat (4.3kJ) must be due to some difference between the lithium ions and sodium ions. This difference is attributed to the fact that the interactions of lithium ions with water produce more heat than the interaction of lithium ions with chloride ions (ion–dipole > ion –ion interaction) whereas the interaction of sodium ions with either water or chloride ions produce about equal enthalpy change.

Problem 52:

Using the data :

$\Delta H^o_f (C_2H_6) = -84.5$ kJ

$\Delta H^o_f (C_2H_4) = 52.6$ kJ

$\Delta H^o_f (C_2H_2) = 226.9$ kJ

$\Delta H^o_f (CO_3CHO) = -166.3$ kJ

$\Delta H^o_f (H_2O,g) = -241.8$ kJ

$\Delta H^o_f (CH_3OH,g) = -201.3$ kJ

$\frac{1}{2}\ DH^o(O_2) = 249.17$ kJ

$\frac{1}{2}$DH° (H – H) = 217.97 kJ

ΔH°C (graphite) → C(g) = 716.68 kJ

calculate the following bond energies:

(i) DH° (C –C), (ii) DH°(C = C),

(iii) DH° (C ≡ C), (iv) DH° (O –H)

(v) DH° (C = O) (vi) DH° (C –O)

Solution:

C–C bond energy. The energy of a C–C bond can be calculated by applying the method of *previous* section to ethane. Our aim is

C_2H_6(g) → 2C(g) + 6H(g) ΔH° = ?

Given data

(i) C_2H_6 (g) → 2C (graphite) + 3H_2 (g) ΔH° = 84.5 kJ

(ii) 2C (graphite) → 2C (g) ΔH° = 1433.4 kJ

(iii) 3H_2 (g) → 6H (g) ΔH° = 1307.8 kJ

Add----------------------------------

C_2H_6 (g) → 2C(g) + 6 H(g) ΔH° = 2825.7 kJ

Thus 2825.7 kJ is required to break 6C – H bonds and one C–C bond in ethane. The C–C bond energy is equal to 2825.7 kJ less 6 × DH° (C –H)

DH° (C – C) = ΔH° – 6 × DH° (C–H)

= 2825.7 –6 × 413 = 347.7 kJ

C = C bond energy. The of C = C bond can be calculated using the standard enthalpy of formation of ethylene and C– H bond energy data. Here our object would be to study :

C_2H_4 (g) → 2C (g) + 4H (g) ΔH° = ?

Given data

C_2H_4 (g) → 2C (graphite)+ 2H_2 (g) ΔH° = – 52. 6 kJ

2C (graphite) → 2C (g) ΔH° = 1433.4 kJ

2H_2(g) → 4H(g) ΔH° = 871.9 kJ

Add ------- ---------

$C_2H_4(g) \rightarrow 2C(g) \quad + 4H(g) \qquad \Delta H° = 2252.7$ kJ

Less 4C – H bonds $\Delta H° = 1652.0$ kJ

-------- --------

C = C bond DH° (C = C) = **600.7 kJ**

The literature value is slightly greater than this value.

C ° C *bond energy*. Since we already know the C– H bond energy, we can obtain the energy of a C ° C bond by using the $\Delta H°_f$ acetylene and the previous C – H bond energy

C_2H_2 (g) → 2C (graphite) + H_2 (g) ΔH° = – 226.9 kJ

2C (graphite) → 2C (g) ΔH° = 1433.4 kJ

H_2 (g) → 2H (g) ΔH° = 435.94 kJ

Add ---------------- -----------

C_2H_2 (g) → 2C (g) + 2H(g) ΔH° = 1642.44 kJ

Less 2C – H bonds 2 × DH° (C– H) = 826.0 kJ

-------- --------

C ≡ C bond ΔH° (C ≡ C) = **816.44 kJ**

C = 0 bond energy is calculated using the standard heat of formation of acetaldehyde and the C– C and C – H bond energies. This gives C = O bond energy as 740 kJ mol^{-1}.

The O – H bond energy

H_2O (g) → $\frac{1}{2}$ O_2 (g) + H_2(g) ΔH° = 241.80 kJ

H_2 (g) → 2H (g) ΔH° = 435.94 kJ

$\frac{1}{2}$ O_2 (g) → O (g) ΔH° = 249.17kJ

Add ------- ----------

H_2O (g) → 2H (g) + O (g) ΔH° = **926.91 kJ**

The quantity, 926.91 kJ is the enthalpy change when the two O – H bonds in H_2O are broken. Therefore, the O – H bond energy is 926.91 kJ/2 = **463.4 kJ.**

Finally, by taking methanol, in which the only unknown bond energy is that of the C–O bond we find that the C– O bond energy is of the order of 351 kJ.

DH° values for different bonds are given in the Appendix.

In the calculation of bond energies we have assumed that one type of bond energy calculated for a particular molecule is the same in all other types of molecules. For example, C–H bond energy in methane, ethane, propane etc., will have the same value. Similarly C–C bond energy for ethane is applicable for C–C bond energy in other molecules too.

However, this procedure has got some limitations, e.g., butane exists in two forms as n-butane and iso-butane. Both compounds have three C–C bonds and ten C– H bonds, and from simple bond theory the heats of formation should be the same for both compounds.

Bond energy data = (3C – C + 10 C–H)

$= 3 \times 348 + 10 \times 413 = 5174.0$ kJ

Heat of atomization = Heat of atomization for (4C + 10H)

$= 4 \times 716.7 + 10 \times 217.97 = 5046.5$ kJ

Standard heats of formation of butane from its elements:

ΔH^{o}_{f} = (Heats of atomization) – (Bond energies)

$= 5046.5 - 5174.0 = -127.5$ kJ

But the observed Δ H°f for n-butane is – 124.3 kJ and that for isobutane is – 131.2 kJ. There is a difference of about 3 kJ in calculated and observed data. This difference becomes more and more as we consider the still higher aliphatic hydrocarbons, *e.g.*, there is a difference of almost 16 kJ in the standard heats of formations of n-pentane and 2-2 dimethyl propane, although both 4 C – C bonds and 12 C – H bonds. Apart from the limitations of the bond energy calculations there are certain merits of the procedure in the sense that it puts restrictions on writing incorrect structures of molecules and suggests a suitable form of representing the molecular structures. We shall discuss the merits of thermodynamic calculations by considering the case of benzene.

Structure of Benzene – Resonating structure – Delocalization concept:

We can calculate the standard heats of formation (DH°f) of benzene in the same manner as that for the simpler hydrocarbons. Usually we write the structure of benzene as a ring with three C– C bonds, three C– C bonds and 6 C – H bonds as

Bond energy data

$= 6 \times DH^\circ (C-H) + 3 \times DH^\circ (C-C) + 3 \times DH^\circ (C=C)$

$= (6 \times 413 + 3 \times 348 + 3 \times 610) = 5352\ mol^{-1}$

Heat of atomization

$= 6 \times \Delta H\ [C\ (graphite) \rightarrow C\ (g)] + 6 \times \Delta H\ (1/2\ H_2 \rightarrow H)$

$= 6 \times 716.7\ kJ\ mol^{-1} + 6 \times 217.96\ kJ\ mol^{-1}$

$= 5608\ kJ\ mol^{-1}$

$\Delta H^\circ f = 5608 - 5352 = 256\ mol^{-1}$

The observed ΔH°_f is only 82.9 kJ mol^{-1}. This shows that benzene molecule is 173 kJ mol^{-1} more stable than predicated by the considered structure. This unexpected stability suggests that the Kekule structure is not a true representative of molecular structure. The real state can be described by a resonance hybrid of the two extreme structures.

The two extreme forms are called "resonance" structures and the difference of the energy of approximately 173 kJ mol–1 is referred to as the resonance stabilization energy. This suggests that the C– C bonds in the benzene ring are neither pure C – C bonds nor pure C = C bonds. All are equivalent and intermediate between single and double bonds. This is shown from C to C bond energy considerations.

Let us suppose that carbon to carbon bonds in the benzene ring are neither pure single nor pure double bonds. Using this approach we can calculate the bond energy of the C linkages.

$C_6H_6 \rightarrow 6C\ (graphite) + 3H_2\ (g) \qquad \Delta H = -\Delta H^\circ_f = -82.9\ kJ$

$6C\ (graphite) \rightarrow 6C\ (g) \qquad \Delta H = 4300.2\ kJ$

$3H_2\ (g) \rightarrow 6H\ (g) \qquad \Delta H = 1307.8\ kJ$

---------- ---------

$C_6H_6 \rightarrow 6C\ (g) + 6H(g) \qquad \Delta H = 5525.1\ kJ$

Thus, the energy required to break six C – H bonds and six C to C bond is 5525.1kJ. But the energy required to break six C– H bonds is 2478 kJ. Therefore, the energy of six C to C linkage would be equal to the difference of the two, i.e., (5525.1 –2478.0) = 3047.1 kJ. Each C to C bond in the ring can be assigned a value of 507.8 kJ. Since the energy of pure C – C bond is 348 kJ and for a pure C = C bond the energy is 610 kJ. This gives an idea that the C to C bonds in benzene have a bond order intermediate between 1 and 2. The bonds between carbon are not localized as double bonds in three places but are spread among all six C– C distances. The electrons involved are delocalized. The extra stability of the molecule arises from the delocalization of the π-electrons in the ring.

Bond energy data are useful in deciding the nature of bonds in a group with multiple bonds. For example, the comparison of bond energies of C – C and C = C shows that both the bonds in C = C are not identical. If this were so then the C = C bond energy must be just double of the C – C bond energy. However, the valency considerations show that C–C bond is a stable bond of σ-type. Of the two bonds in C = C linkage one is s-type and the other is a weaker p-linkage. This is in line with the thermodynamic calculations which gives a difference of 262 kJ mole–1 between C = C bond and C – C bond energies. This suggests that the bond other than σ-type in C = C bond is weaker. Similarly in C ≡ C linkage the nature of all the three bonds are not identical, one being s-type and the two are π-type bonds.

Problem 53:

Using the bond energy data, compute ΔH^o_f for ethyl alcohol (C_2H_5OH). DH° (C – H) = 413 kJ, D H° (C – C) = 348 kJ, DH° (C – O) = 351 kJ, DH° (O – H) =463kJ.

Solution :

Ethyl alcohol can be represented by the structure

```
     H    H
     |    |
H— C —  C — O — H
     |    |
     H    H
```

Total bond enthalpy = 5 × DH° (C – H) + DH° (C–C)+DH°(C–O) + DH°(O –H)

= 5 × (413 kJ) + 348 kJ + 351 kJ + 363 kJ = 3227 kJ

when 3227 kJ is supplied then 2C (g), 6H (g) and O (g) will be the product given by C_2H_5OH.

Total heat of atomization = 6H (g) + 2C (g) + O (g)

= 6 × 21.797 + 2 × 716.68 + 249.2 = 2990.38 kJ

$\Delta H^o f$ (C_2H_5OH) = Enthalpy of atomization – Bond enthalpy
= (2990.38 kJ) – (3227.0 kJ) = – **236.62 kJ**

Heats of Reaction from Bond Energy Data

Bond energy data are used to calculate the heats of gaseous reactions involving substances having only covalent bonds. In using these data, a plus sign in affixed to the bond enthalpies when a bond is broken, since heat is absorbed under such conditions and a minus sign when a bond is formed and heat is evolved.

Problem 54:

Using bond energy data calculate the enthalpy change for the reaction:

C_2H_4 (g) + H_2 (g) → C_2H_6 (g)

Solution:

The given reaction can be written as

```
H    H           H   H
|    |           |   |
C =  C + H – H → H – C – C – H
|    |           |   |
H    H           H   H
```

In this reaction the four C – H bonds in C_2H_4 are unaffected and a double bond is broken ; in H_2 one H – H bond is broken. In turn, two new C – H bonds and one C – C bond are formed in ethane.

ΔH^o_f = Bond enthalpy in – Bond enthalpy out

= DH° (C = C) + DH° (H – H) 2 × D H° (C – H) – DH° (C – C)

= (610 kJ mol^{-1}) + (436 kJ mol^{-1}) –2 × (413 kJ mol^{-1})

– (348 kJ mol^{-1})

= **–128** kJ mol^{-1}.

Problem 55:

Calculate the standard enthalpy of formation of dimethyl ether by using the bond energy data from the appendix and the enthalpy of C (graphite) → C (gas) as 716.7 kJ.

Solution:

The formation of dimethylether may be represented by the equation

(A) 2C (graphite) + $3H_2(g)$ + 1/2 O_2 (g) → $H_3C–O–CH_3$ ΔH_f = ?

From bond energy data we get the enthalpy for the following reaction:

(i) $H_3C – O – CH_3$ (g) → 2C (g) + 6H (g) + O(g)

$\Delta H = 6DH° (C – H) + 2D H° (C – O)$

$\therefore \Delta H = 6 \times (413 \text{ kJ mol}^{-1}) + 2 \times (350 \text{ kJ mol}^{-1}) = 3178 \text{ kJ}$

(ii) $3H_2$ (g) → 2H (g) ΔH = 1307.64 kJ

(iii) 1/2 O_2 (g) → O (g) ΔH = 249.17 kJ

(iv) 2C (graphite) → 2C (g) ΔH = 1433.4 kJ

Reversing eqn.. (i) we get

(v) 2C (g) + 6H (g) + O(g) → $H_3C–O–CH_3$ (g) ΔH = – 3178 kJ

Adding equations (ii), (iii), (iv) and (v), and cancelling the common terms we get

(iv) 2C (graphite) + $3H_2$ (g) + 1/2 O_2 (g) → $H_3 C – O – CH_3$ (g)

ΔH (vi) = ΔH (ii) + ΔH (iii) + ΔH (iv) + ΔH (v)

= – 189.79 kJ = DH (A).

2

THERMODYNAMICS PROCESS AND FUNCTIONS

INTRODUCTION

The operation which brings about the changes in the state of the system is called the process. For example, heating is the process in the above example, similarly the various operations like vaporization, freezing, sublimation expansion, contraction by which the state of the system can be changed, are known as processes. We can discuss below:

Isothermal process : When the temperature of the system is kept constant during various operations then the process is said to be isothermal. This is attained either by removing heat from the system or by supplying heat to the system. For isothermal process $dT = 0$.

Adiabatic process : There is no exchange of heat between the system and surroundings and q is equal to zero. Thus is adiabatic processes there is always a change in a temperature.

Isobaric process: When the pressure on the system remains constant during the period of the change, the process is said to be isobaric. For an isobaric process $dP = 0$. For example, conversion of 1 mole of water at 25° C and 1 atm pressure into vapour at 100°C and 1 atm pressure is isobaric process.

Isochoric process : When there is no change in the volume of the system during various operations the change is said to be isochoric. For such changes $dV = 0$.Combustion of a substance in a bomb calorimeter is an isochoric process.

Cyclic process : If a system having undergone a change returns to initial state then the process is called a cyclic process and the path of the process is called a cycle.

Path : Sequence of steps starting from the initial state to the various intermediate states and ultimately to the final state of the system is called the path. Thus the path of conversion of 1 mole of water at 25°C and 1 atm into vapour at 100°C abd 1 atm can be represented as

$$\underset{\text{(Initial state)}}{1 \text{ mole } H_2O \text{ (liq., } 25° \text{ C, 1 atm)}} \xrightarrow{\text{Heat}} \underset{\text{(a)}}{H_2O \text{ (liq., } 50°\text{C, 1 atm)}} \xrightarrow{\text{Heat}}$$

$$\underset{\text{(b)}}{H_2O \text{ (liq., } 75°\text{C, 1 atm)}} \xrightarrow{\text{Heat}} \underset{\text{(c)}}{H_2O \text{ (liq., } 100° \text{ C, 1 atm)}}$$

$$\underset{\text{(Final state)}}{H_2O \text{ (vapour, } 100°\text{C, 1 atm)}}$$

(a), (b), (c) etc. are the intermediate states and whole of the scheme represents the path of the system.

The final state of a system can be attained either directly or in stages. Accordingly there are two types of paths of a system, consequently two types of processes are recognized. These are :

Reversible process : When the process occurs in such a way that the properties of the system at every instant remain uniform, the process is called reversible. Thus, reversible process is one which proceeds through a succession of equilibrium steps, each of which is an equilibrium state. A reversible process must be carried out at a very slow rate so that enough time is allowed between each change for the properties to come to a constant value. Reversible processes are ideal. In actual practice it is not possible always to perform the experiments under such an ideal situation. Nevertheless the concept of reversibility serves as a means to study the very many processes which are of immense importance to chemists.

Irreversible process : If the change is produced rapidly and the system does not get a chance to attain equilibrium then the process is called irreversible. All natural processes are irreversible and hence *spontaneous*. Once undergone a change the process cannot be reverted without the help of external source and without changing the properties of the surroundings.

Examples of a reversible process are :

(i) Vaporization of a liquid in a closed vessel at a constant temperature is a reversible process. At this temperature the equilibrium pressure of the vapour is equal to the saturation vapour pressure.

Now, if the pressure is increased by any amount, the vapour will condense. On the other hand when the pressure is diminished by an infinitesimal amount compared to the equilibrium vapour pressure, the liquid will vaporise. It has to be noted that this does not bring about any change in the surroundings.

(ii) Chemical reaction in a galvanic cell is very nearly a reversible process. The electromotive force (emf) of a cell is measured by a sensitive potentiometer when the driving force of the cell is balanced against an external emf, and the current can be made to flow in one direction or the other by external change of one micro volt without leaving more than a vanishingly small change in the surroundings.

(iii) Chemical reactions of the type A B also come under the class of reversible process.

Examples of a few irreversible processes are – expansion of a gas into vacuum, reaction for the formation of curd from milk, in fact all natural processes are irreversible.

Let us illustrate the significance of the concepts of reversible and irreversible processes for a gas contained in a cylinder which is fitted with a piston. Let the gas in the cylinder comprise the system and let the piston be loaded with a large number of small weights as shown in Fig. 1. Let the weights be slid off horizontally one at a time, allowing the gas to expand and to work in raising the weights which remain on the piston. As the size of the weights is small and their number is large, the expansion would be reversible for at each level of piston during the reverse process there will be small weight which is exactly at the level of the platform and thus can be placed on the platform without requiring work. In the limit, therefore, as the weights become very small, the reverse process can be accomplished in such a manner that both the system and surrounding are exactly in the same state they were initially.

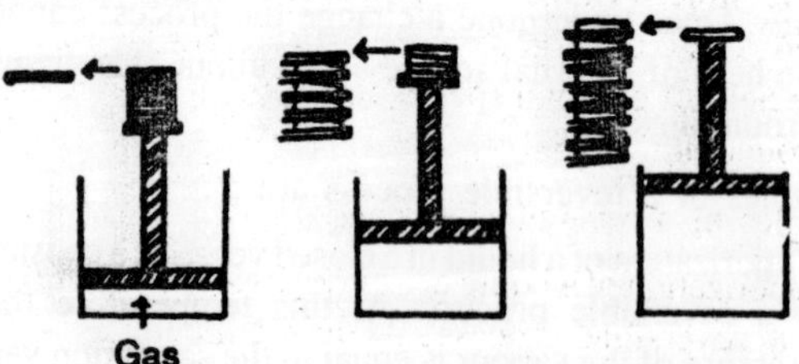

Fig. 1 : An Example of a reversible process.

Let us consider another situation. Let the gas in the cylinder be at high pressure restrained by a piston which is secured by pin s depicted in Fig. 2. When the pin is removed, the piston is raised and forced abruptly against the stops.

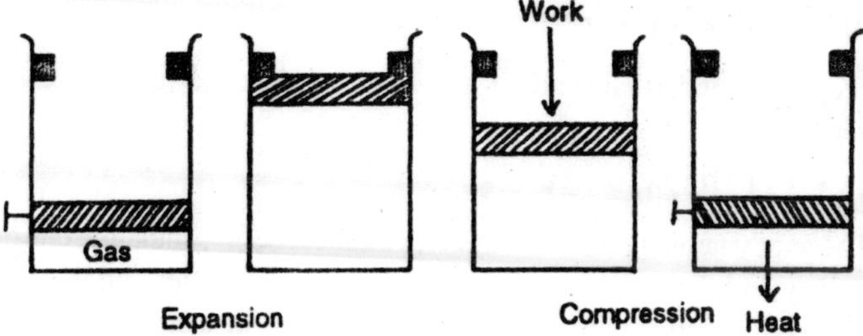

Fig. 2 : An example of irreversible process.

Some work is done by the system (gas in he cylinder), since the piston has been raised. If we wish to restore the system to its initial state, a force must be exerted on the piston to compress the gas until the pin could again be inserted in the piston. Since the work done on the gas during compression is greater than the work done by the gas in the expansion process, an amount of heat must be transferred from the system to the surroundings in order that the system is at the same initial state.

Thus the system is restored to its initial state, but the surroundings have changed by virtue of the fact that work was required to force the piston down and heat was transferred to surroundings. Thus the expansion of gas under the present condition is an irreversible process because it could not be reversed without leaving a change in the surroundings.

HOMOGENEOUS FUNCTION

Any function f (x, y, z) is said to be homogeneous of degree m of the variables x, y, z, if

$$f(\lambda x, \lambda y, \lambda z) = \lambda^m f(x, y, z) \quad ...(1)$$

for all positive values of l. This identity implies that an increase in all variables by a factor λ produces an increase in the function by a factor of λ^m. If the degree of homogeneity is unity (m = 1), the operation of multiplying each of the independent variables by a parameter λ produces the same effect as multiplying the overall function by the same factor.

All the extensive properties of a system are homogeneous of first degree of moles. Volume is an extensive property and also it is a homogeneous function of first degree of mole, because when we double the moles of the substance its volume is also doubled.

That is for

$$V = f(T, P, n)$$

keeping T and P constant if n is doubled then the new volume V′ is given by,

$$V' = f(T, P, 2n) = 2V$$

Thermodynamic functions namely, volume, energy, enthalpy, entropy, free energy and the work function are extensive properties and hence are homogeneous function of the first degree in moles. The intensive properties are homogeneous function of zeroth degree of moles. For example, pressure is an intensive property and hence a homogeneous function of zeroth degree *i.e.*

$$P(\lambda n_1, \lambda n_2 ...) = \lambda^{\circ} P(n_1, n_2 ...)$$

Euler's Theorem on Homogeneous Functions

This theorem states that if the function f (x, y, z) is a homogeneous function of degree m then

$$x\frac{\partial f}{\partial x} + y\frac{\partial f}{\partial y} + z\frac{\partial f}{\partial z} = mf(x, y, z) \quad ...(2)$$

and for a first degree function (m = 1), we have

$$x\frac{\partial f}{\partial x} + y\frac{\partial f}{\partial y} + z\frac{\partial f}{\partial z} = f(x, y, z) \quad ...(3)$$

These equations can be generalised for any number of independent variables. This theorem is very useful to deduce the relations involving partial molar quantities.

PROPERTIES OF STATE FUNCTIONS–EXACT DIFFERENTIALS

State functions give exact differentials. These exact differentials can be integrated between limits, without regard for the actual change that occur as the system moves from one limiting condition to the other. The properties of a state function can be understood if we take a concrete example of volume of definite amount of the substance. It depends upon temperature (T) and pressure (P). Mathematically we can speak that volume is a function of temperature and pressure *i.e.*,

$V = f(T, P)$ (for a definite amount)

If T and P are given a particular value of V is thereupon fixed. If T and P are changed the value of V will accordingly be changed. For this reason T and P are called *independent variables* and V is called the *dependent variable.*

The change in V can be estimated provided the derivatives of the function V with respect to T and P are known. The derivative is the rate of change of dependent variable with the independent variable. For the present case the two derivatives are :

$(\partial V/\partial T)_P$ = rate of change of volume with temperature at constant pressure.

$(\partial V/\partial P)_T$ = rate of change of volume with, pressure at constant temperature.

If the derivative is positive the function V increases with the increase of the independent variable; if the derivative is negative the function decreases with the increase of the independent variable. If the derivative is zero the function is independent of the variables, and the value of the function may be maximum or minimum under that condition.

The change of volume due to infinitesimal change of temperature (dT) at constant pressure is equal to the rate of change of volume with temperature $(\partial V/\partial T)_P$ multiplied by dT, and is given by

$$dV = (\partial V/\partial T)_P \, dT$$

The change in volume due to infinitesimal change of pressure (dP) at constant temperature is given by

$$dV = (\partial V/\partial P)_T \, dP$$

If both the temperature and pressure are changed simultaneously then the total change in volume (dV) is given by

$$dV = \left(\frac{\partial V}{\partial T}\right)_P dT + \left(\frac{\partial V}{\partial P}\right)_T dP$$

This is called the total differential of the function (V). This equation was obtained by assuming that *whether we first change temperature then pressure or, first change the pressure then temperature, the final total change in volume is the same*. This is true because V is a *state function* and dV is *an exact differential*. In the language of mathematics the condition of exactness is

$$\frac{\partial}{\partial P}\left[\left(\frac{\partial V}{\partial T}\right)_P\right]_T = \frac{\partial}{\partial T}\left[\left(\frac{\partial V}{\partial P}\right)_T\right]_P$$

or $$\frac{\partial^2 V}{\partial P \partial T} = \frac{\partial^2 V}{\partial T \partial P}$$

This relation is applicable for any state function ϕ.

Let $\phi = f(x, y)$, then total differential $d\phi$ is given by

$$d\phi = \left(\frac{\partial \phi}{\partial x}\right)_y dx + \left(\frac{\partial \phi}{\partial y}\right)_x dy = M(x,y)\,dx + N(x, y)\,dy$$

where $M(x, y) = \left(\frac{\partial \phi}{\partial x}\right)_y$ and $N(x, y) = \left(\frac{\partial \phi}{\partial y}\right)_x$ ...(1)

If $d\phi$ is an exact differential then

$$\left(\frac{\partial M}{\partial y}\right)_x = \left(\frac{\partial N}{\partial x}\right)_y \quad ...(2)$$

$$\therefore \frac{\partial^2 \phi}{\partial y dx} = \frac{\partial^2 \phi}{\partial x \partial y} \quad ...(3)$$

Here ϕ can be considered as any thermodynamic function (E, H, G etc). and (x, y) may be thermodynamic T, P, V etc. The relations given by equations (1, 2) and (1.3) are called the Euler's theorem of exactness. Any function and dY is inexact differential.

Some other properties of derivatives :

(i) They can be turned upside down, the subscript remaining the same.

$$1\Big/\left(\frac{\partial \phi}{\partial x}\right)_y = \left(\frac{\partial x}{\partial \phi}\right)_y$$

However, $$1\Big/\left(\frac{\partial^2 \phi}{\partial x^2}\right) \neq \frac{\partial x^2}{\partial^2 \phi}$$

(ii) They can be multiplied together cancelling numerator with denominator provided the subscripts are the same.

$$\left(\frac{\partial \phi}{\partial x}\right)_y \left(\frac{\partial x}{\partial z}\right)_y = \left(\frac{\partial \phi}{\partial z}\right)_y$$

But $\left(\frac{\partial\phi}{\partial x}\right)_y \left(\frac{\partial x}{\partial z}\right)_t \neq \left(\frac{\partial\phi}{\partial z}\right)$

ENTROPY

The entropy natural processes are accompanied with an increase in the total entropy of the system and its surroundings. This gives us the equilibrium condition when the system cannot undergo any spontaneous change. At this state the entropy of the isolated system is maximum. Mathematically it can be explained with the help of the combined equation of the first and second laws of thermodynamics:

(a) $dS = \frac{dE + PdV}{T}$ for a reversible process

But $dE = C_V\ dT + (\partial E/\partial V)_T\ dV$

$\therefore\ dS = C_V\ \frac{dT}{T} + \frac{1}{T}[(\partial E/\partial V)_T + P]dV$

In an isolated system dE = 0, dT = 0 any dV = 0, therefore

$(dS)_{E,V} = 0$

For an irreversible change, since $(PdV)_{rev} > (PdV)_{irr}$,

$\therefore\ dS > \frac{dE + PdV_{irr}}{T}$

In an isolated system, dE = 0, dV = 0 and hence

$(dS)_{E,V} > 0$ ...(1)

The combined result may be written as

$(dS)_{E,V} \geq 0$

The state of equilibrium correspond of maximum entropy of the universe.

(b) If the entropy and volume are constant, then dS = 0 dV, and we should have

$(dE)_{S,V} = 0$ for a reversible change.

For an irreversible change $TdS > dE + PdV_{irr}$ or

$dE < TdS - PdV_{irr}$

$\therefore\ (dE)_{S,V} < 0$ ...(2)

The system maintained at constant entropy and constant volume will attain the equilibrium at a state of minimum energy.

(c) At constant entropy and pressure another condition of equilibrium in terms of enthalpy may be given. We have the relation dH = TdS – VdP, at constant S and P, dS = 0, dP = 0

$\therefore (dH)_{S,P} = 0$

For irreversible change $(dH)_{S,P} < 0$.

Thus the state of equilibrium is attained when the energy is minimum and the entropy is maximum. These are the necessary conditions for equilibrium. The sufficient condition of equilibrium is that if any virtual displacement (δ) is made in the variables of the system, then entropy in the new state will be less than the old equilibrium state and $\delta S < 0$, and the energy in the new state will be larger than the old equilibrium state and $\delta E > 0$. This can be understood by considering the Fig. 3(a) and 3(b) for a process A $(T_1, V_2, E_1, S_1) \rightarrow (T_2, V_2, E_2, S_2)$.

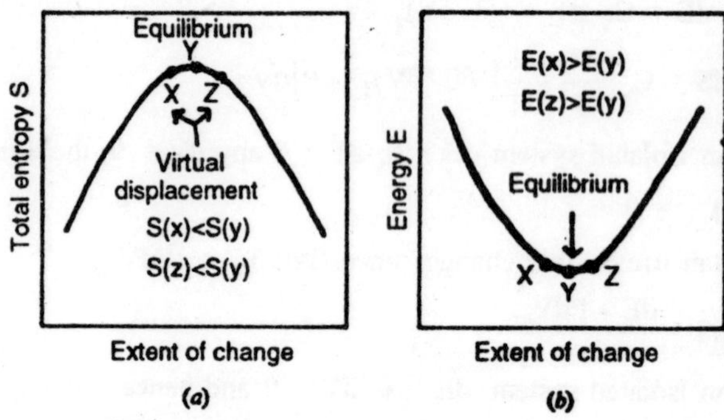

Fig. 3 : The condition for equilibrium (a) The entropy is maximum an $\delta S < S < 0$; (b) The energy is minimum and dE > 0.

When equilibrium is attained, the entropy is maximum and energy is minimum. Now suppose that by some means, the equilibrium is displaced to new states X or Z. From the figure it is quite clear that the entropy in the state X is less than the entropy in the state Y i.e. S (X) < S(Y) and δS in negative.

Similarly S (Z) < S (Y) and $\delta S < 0$. The energy in either of the displaced state is greater than the energy at equilibrium i.e. E (X) > E (Y) and E (Z) > E (Y), therefore $\delta E > 0$ is the sufficient condition for equilibrium.

Necessary condition . $(dS)_{E,V} > 0$; $(dE)_{S,V} < 0$.

Sufficient condition. $\delta S < 0$; $\delta E > 0$.

ENTROPY CHANGES SYSTEMS

In calculation of entropy changes we must take into consideration the entropy of system and the surroundings. The combination of system and surroundings correspond to an isolated system. The surroundings are usually considered to be the (i) heat reservoir that can add to, or subtract heat from the system, or (ii) mechanical device which can do work on, or accept work from the system. Thus the total entropy change of the isolated system ($\Delta S_{universe}$) is taken as

$$\Delta S_{universe} = \Delta S_{system} + \Delta S_{surroundings}$$

Entropy Change of System (ΔS_{system})

Entropy change of the system in a reversible isothermal change can be calculated by using the relation $dS = dq_{rev}/T$. But this equation cannot be used as such to calculate the entropy of irreversible changes. For this we have to devise the ways to carry the change in reversible ways. This is valid because the entropy is a state function and the changes in entropy of the system are the same irrespective the way the change is brought about.

Entropy Change of the Surroundings ($\Delta S_{surroundings}$)

The entropy changes for the surroundings can be calculated if we know the gain or loss in the heat of the surroundings. For this case, the equation $dS = dq/T$ can be used to calculate $\Delta S_{surroundings}$ because the surroundings are so large that the removal or addition of heat can be taken as reversible processes, as it will not alter the properties of the surroundings to any greater extent. [For analogy we take the case that removal or addition of a few drops of water will not change the water content of ocean, (Surroundings)].

$$\Delta S_{surroundings} = -\frac{q_{system}}{T} = \frac{q_{surroundings}}{T}$$

Now we shall consider the calculation of entropy change for some processes.

Phase changes. As an example of phase changes we may consider the melting of a solid. At melting point, the liquid and solid exist together in equilibrium. The addition of a little heat would melt some of the solid, the removal of a little heat would solidify some of the liquid, but the solid liquid equilibrium will be maintained. This process can be considered as reversible and entropy change can be given by

$$\Delta S_f = \int_1^2 \frac{dq_{rev}}{T_f}$$

Since the temperature is constant during the process, it may be taken outside the integral, giving

$$\Delta S_f = \frac{1}{T_f} \int dq_{rev}$$

The integral is the total heat absorbed which for this constant pressure process is the enthalpy of fusion, so that

$$\Delta S_f = \frac{\Delta H_f}{T_f} \qquad ...(1)$$

In general, for any phase transformation taking place at constant temperature and pressure, e.g. [solid $\rightleftharpoons$ liquid (melting); liquid $\rightleftharpoons$ vapour (vaporization); solid $\rightleftharpoons$ vapour (sublimation); solid (α) $\rightleftharpoons$ solid (β) (transition) it must be true that

$$\Delta S_t = \frac{\Delta H_t}{T_t} \qquad ...(2)$$

where ΔH_t is the enthalpy change of the transformation and T_t is the constant temperature at which the transformation takes place.

THERMAL EXPANSION (α)

The rate of change of volume with temperature at constant pressure relative to a volume V is thermal expansively α.

$$\alpha = \frac{1}{V}\left(\frac{\partial V}{\partial T}\right)_P$$

a has the dimensions of T^{-1}.

Compressibility Coefficient (β)

The rate of change of volume with pressure at constant temperature relative to a volume is defined as the compressibility of a substance.

$$\beta = -\frac{1}{V}\left(\frac{\partial V}{\partial P}\right)_T$$

The dimensions of β are those of P^{-1}. The expression for b bears a minus sign; this stands for the fact that volume decreases with increase of pressure and *vice versa*.

Relation between α and β

$V = f(T, P)$

$$dV = \left(\frac{\partial V}{\partial T}\right)_P dT + \left(\frac{\partial V}{\partial P}\right)_T dP$$

or $dV = \alpha V\ dT - \beta V\ dP$ ...(1)

For a condition of constant volume, $dV = 0$

$\therefore\ \alpha V(\partial T)_V - \beta V - \beta V\ (\partial P)_V = 0$

$$\text{or } \left(\frac{\partial P}{\partial T}\right)_V = \frac{\alpha}{\beta}$$

At constant pressure $dP = 0$ and equation (1) reduces to

$$\left(\frac{\partial V}{\partial T}\right)_P = \alpha V$$

Differentiating it with respect to P at constant T, we get

$$\frac{\partial}{\partial P}\left[\left(\frac{\partial V}{\partial T}\right)_P\right]_T = \alpha\left(\frac{\partial V}{\partial P}\right)_T + V\left(\frac{\partial \alpha}{\partial P}\right)_T$$

$$= -\alpha V\beta + V\left(\frac{\partial \alpha}{\partial P}\right)_T \qquad ...(2)$$

At constant temperature $dT = 0$ and equation (1) transforms to

$$\left(\frac{\partial V}{\partial P}\right)_T = -\beta V$$

Differentiating with respect to T at constant to T constant P we obtain

$$\frac{\partial}{\partial T}\left[\left(\frac{\partial V}{\partial P}\right)_T\right]_P = -\beta\left(\frac{\partial V}{\partial T}\right)_P - V\left(\frac{\partial \beta}{\partial T}\right)_P$$

$$= -\beta V\alpha - V\left(\frac{\partial \beta}{\partial T}\right)_P \qquad ...(10)$$

Since V is a state function and $\partial^2 V/\partial T\ \partial P = \partial^2 V/\partial P\ \partial T$, therefore from equations (2) and (3) we obtain

$$V\left(\frac{\partial \alpha}{\partial P}\right)_T = \left(\frac{\partial \beta}{\partial T}\right)_P$$

$$\text{or} \quad \left(\frac{\partial \alpha}{\partial P}\right)_T + \left(\frac{\partial \beta}{\partial T}\right)_P = 0 \qquad ...(4)$$

SOME SOLVED PROBLEMS

Problem 1:

If $E = f(T, V)$ and dE is an exact differential then show that $\left(\frac{\partial E}{\partial V}\right)_T = T\left(\frac{\partial P}{\partial T}\right)_V - P$, *relationship can be derived by using the relation $dq = dE + PdV$ and taking $(1/T)$ as an integrating factor.*

Solution:

Since E = f (T, V)

$$\text{Therefore, } dE = \left(\frac{\partial E}{\partial T}\right)_V dT + \left(\frac{\partial E}{\partial V}\right)_T dV$$

Given that dq = dE + PdV

$$\text{Now } dE = \left(\frac{\partial E}{\partial T}\right)_V dT + \left(\frac{\partial E}{\partial V}\right)_T dV$$

$$\therefore dq = \left(\frac{\partial E}{\partial T}\right)_V dT + \left(\frac{\partial E}{\partial V}\right)_T + P\Bigg] dV$$

Let $f = \frac{1}{T}$ be the integrating factor, then f dq must be an exact differential.

$$fdq = \frac{1}{T}\left(\frac{\partial E}{\partial T}\right)_V dT + \frac{1}{T}\left[\left(\frac{\partial E}{\partial V}\right)_T + P\right] dV$$

$$= M\,dT + N\,dV$$

$$\text{where, } M = \frac{1}{T}\left(\frac{\partial E}{\partial T}\right)_V ;\ N = \frac{1}{T}\left[\left(\frac{\partial E}{\partial T}\right)_T + P\right]$$

For the above differential to be exact we must have

$$\left(\frac{\partial M}{\partial V}\right)_T = \left(\frac{\partial N}{\partial T}\right)_V$$

or $$\frac{\partial}{\partial V}\left[\frac{1}{T}\left(\frac{\partial E}{\partial T}\right)_V\right]_T = \frac{\partial}{\partial T}\left[\frac{1}{T}\left(\frac{\partial E}{\partial T}\right)_T + \frac{P}{T}\right]_V$$

or $$\frac{1}{T}\frac{\partial^2 E}{\partial V \partial T} = \frac{1}{T}\frac{\partial^2 E}{\partial T \partial V} - \frac{1}{T^2}\left(\frac{\partial E}{\partial V}\right)_T - \frac{P}{T^2} + \frac{1}{T}\left(\frac{\partial P}{\partial T}\right)_V \qquad \text{...(i)}$$

Since dE is an exact differential, therefore,

$$\frac{\partial^2 E}{\partial T \partial V} = \frac{\partial^2 E}{\partial V \partial T} \qquad \text{...(ii)}$$

From equations (i) and (ii), we get

$$-\frac{1}{T^2}\left(\frac{\partial E}{\partial V}\right)_T - \frac{P}{T^2} + \frac{1}{T}\left(\frac{\partial P}{\partial T}\right)_V = 0$$

or $$\left(\frac{\partial E}{\partial V}\right)_T = T\left(\frac{\partial P}{\partial T}\right)_V - P$$

Exercise: If H = f (T, P) and dH is an exact differential then prove that

$$\left(\frac{\partial H}{\partial P}\right)_T = V - T\left(\frac{\partial V}{\partial T}\right)_P$$

[**Hint :** Start with dq = dH – V dp taking 1/T as an integrating factor.]

Problem 2:

A gas obeys the equation of state $P = nRT/V - n^2 RB/V^2$. Show that

(i) $\left(\frac{\partial S}{\partial V}\right)_T = nR/V$; (ii) $(\partial E/\partial V)_T = n^2BR/V^2$, (iii) $(\partial H/\partial V)_T = 2n^2BR/V^2$ *where B is a constant.*

Solution:

(i) $(\partial S/\partial V)_T$ suggests that S = S (T, V), therefore,

$$dS = \left(\frac{\partial S}{\partial T}\right)_V dT + \left(\frac{\partial S}{\partial V}\right)_T dV \qquad \text{...(i)}$$

From the first and second laws we get

$$dS = dq_{rev}/T = \frac{1}{T}(dE + PdV) \qquad \text{...(ii)}$$

But E = E (T,V) and hence

$$dE = \left(\frac{\partial E}{\partial T}\right)_V dT + \left(\frac{\partial E}{\partial V}\right)_T dV = C_V dT + \left(\frac{\partial E}{\partial V}\right)_T dV$$

$$\therefore\ dS = \frac{C_V}{T} dT + \frac{1}{T}\left[\left(\frac{\partial E}{\partial V}\right)_T + P\right] dV$$

Comparing the coefficients of dV in equations (i) and (iii) it is seen that

$$\left(\frac{\partial S}{\partial V}\right)_T = \frac{1}{T}\left[\left(\frac{\partial E}{\partial V}\right)_T + P\right]$$

$$\text{But } \left(\frac{\partial E}{\partial V}\right)_T = T\left(\frac{\partial P}{\partial T}\right)_V - P \qquad \text{...(iii)}$$

$$\therefore\ \left(\frac{\partial S}{\partial V}\right)_T = \left(\frac{\partial P}{\partial T}\right)_V$$

For the given gas $(\partial P/\partial T)_V = nR/V$

$$\therefore\ \left(\frac{\partial S}{\partial V}\right)_T = nR/V \qquad \text{...(iv)}$$

(ii) Putting $(\partial P/\partial T)_V = nR/V$ in thermodynamic equation of state we get

$$(\partial E/\partial V)_T = T\frac{nR}{V} - P = n^2RB/V^2 \qquad \text{...(v)}$$

(iii) H = E + PV

$$\therefore\ \left(\frac{\partial H}{\partial V}\right)_T = \left(\frac{\partial E}{\partial V}\right)_T + P + V\left(\frac{\partial P}{\partial V}\right)_T \qquad \text{...(vi)}$$

For the given equation $P = nRT/V - n^2 RB/V^2$

$$\left(\frac{\partial P}{\partial V}\right)_T = -\,nRT/V^2 + 2n^2 RB/V^3$$

$$\therefore\ V\left(\frac{\partial P}{\partial V}\right)_T = -\,nRT/V + 2n^2RB/V^2 \qquad \text{...(vii)}$$

Also $(\partial E/\partial V)_T = n^2 RB/V^2$

On substituting the derivatives in equation (vi) we get

$$\left(\frac{\partial H}{\partial V}\right)_T = \frac{n^2RB}{V^2} + P - \frac{nRT}{V} + P - \frac{2n^2RB}{V^2}$$

$$\text{But } P - \frac{nRT}{V} = \frac{n^2RB}{V^2}$$

$$\therefore \quad \left(\frac{\partial H}{\partial V}\right)_T = \frac{2n^2RB}{V^2}$$

Problem 3:

A gas obeying the equation of state P (V –B) = RT undergoes a change from the initial state T_1, V_1 to a final state T_2, V_2. Derive an expression for the entropy change of this gas. The variation of heat capacity at constant volume is given by $C_V = a + bT + cT^2$.

Solution:

$$dS = \frac{dE + PdV}{T}, \ dE = C_V \, dT + (\partial E/\partial V)_T \, dV \text{ give}$$

$$dS = \frac{C_V}{T} dT + \frac{1}{T}[(\partial E/\partial V)_T + P]dV \quad ...(i)$$

Thermodynamic equation of state $(\partial E/\partial V)_T + P = T\,(\partial P/\partial T)_V$, when applied, equation (i) yields,

$$dS = \frac{C_V}{T} dT + (\partial P/\partial T)_V dV \quad ...(ii)$$

For the given gas P = RT/(V – B), and

$(\partial P/\partial T)V = R/(V - B)$

Substituting equation (iii) into (ii) we obtain

$$dS = \frac{C_V}{T} dT + R\frac{dV}{V-B} \quad ...(iv)$$

$$\because \ C_V = a + bT + cT^2$$

$$\therefore \ \frac{C_V}{T} = \frac{a}{T} + b + cT \quad ...(vi)$$

$$dS = \frac{a}{T} dT + bdT + cTdT + R \frac{dV}{V-B}$$

On integrating for proper limits, we get

$$\int_1^2 dS = a\int_{T_1}^{T_2} \frac{dT}{T} + b\int_{T_1}^{T_2} dT + c\int TdT + R\int_{V_1}^{V_2} \frac{dV}{V-B}$$

$$\therefore \Delta S = a \ln \frac{T_2}{T_1} + b(T_2 - T_1) + \frac{c}{2}(T_2^2 - T_1^2) + R \ln \frac{V_2 - B}{V_1 - B}$$

Problem 4:

(a) *Starting with the fundamental relations prove that*

$$\left(\frac{\partial S}{\partial V}\right)_T = \left(\frac{\partial P}{\partial T}\right)_V = \frac{\alpha}{\beta}, \text{ where } a = V^{-1}(\partial V/\partial T)_P,\ \beta = V^{-1}(\partial V/\partial P)_T$$

Solution:

(a) Combined first and second laws of thermodynamics give

$TdS = dE + PdV$

But $dE = C_V\, dT + \left(\frac{\partial E}{\partial V}\right)_T dV$

$$\therefore TdS = C_V\, dT + \left[\left(\frac{\partial E}{\partial V}\right)_T + P\right] dV \quad \text{...(i)}$$

At constant T, dT = 0 and

$$T\left(\frac{\partial S}{\partial V}\right)_T = \left(\frac{\partial E}{\partial V}\right)_T + P \quad \text{...(ii)}$$

Thermodynamic equation of state reads

$$(\partial E/\partial V)T + P = T\left(\frac{\partial P}{\partial T}\right)_V$$

$$\therefore T\left(\frac{\partial S}{\partial V}\right)_T = T\left(\frac{\partial P}{\partial T}\right)_V$$

$$\text{or} \quad \left(\frac{\partial S}{\partial V}\right)_T = \left(\frac{\partial P}{\partial T}\right)_V \quad \text{...(iv)}$$

Also $V = f(T, P)$

$$\therefore dV = \left(\frac{\partial V}{\partial T}\right)_P dT + \left(\frac{\partial V}{\partial P}\right)_T dP = \alpha V dT - \beta V dP$$

At constant volume dV = 0 and

$$\left(\frac{\partial P}{\partial T}\right)_V = \frac{\alpha}{\beta} \quad \text{...(v)}$$

From equation (iv) and (v), we get

$$\left(\frac{\partial S}{\partial V}\right)_T = \left(\frac{\partial P}{\partial T}\right)_V = \alpha/\beta$$

(b) *For a given substance a = 1.24 × 10–3 K⁻¹ and b = 9.3 × 10⁻⁵ atm–1 at 290 K and 1 atm. Assuming a and b to be constant, calculate the change in molar volume which should produce an entropy change of 2.1 JK⁻¹ mol⁻¹ at 290 K.*

Solution:

(b) We have derived $\left(\frac{\partial S}{\partial V}\right)_T = \left(\frac{\partial P}{\partial T}\right)_V = \alpha/\beta$

or $\frac{\Delta S}{\Delta V} = \alpha/\beta$

$\therefore\ \Delta V = \Delta S\ \beta/\alpha$

$$= \frac{(2.1 JK^{-1} mol^{-1}) \times (9.3 \times 10^{-5} atm^{-1})}{(1.24 \times 10^{-3} K^{-1})}$$

$$= 0.1583\ mol^{-1}\ atm^{-1} = \frac{0.1583 Jmol^{-1} atm^{-1}}{101.325 Jatm^{-1} dm^{-3}}$$

$= 1.56 \times 10^{-3}\ dm^3\ mol^{-1}$

Problem 5:

(a) *Calculate the entropy change involved in the conversion of 1mole of ice at 0°C and 1 atm to liquid at 0° C and 1 atm; the enthalpy of fusion per mole of ice is* 6008 J mol⁻¹.

Solution:

The process is reversible.

H_2O (s, 0°C, 1 atm) → H_2O (l, 0° C, 1 atm)

$$\therefore\ \Delta S_{fus} = \frac{\Delta H_{fus}}{T_m}$$

$T_m = 273\ K;\ \Delta H_{fus} = 6008\ J\ mol^{-1}$

$$\therefore\ \Delta S_{fus} = \frac{6008 J\ mol^{-1}}{273 K} = 22\ \mathbf{JK^{-1}}\ mol^{-1}$$

This is an increase in the entropy of the system. This means that the entropy of liquid state is greater than that of the solid state of H_2O. Since the randomness in liquid is greater than solid, hence entropy is a measure of randomness.

Problem 6:

(b) *Calculate the change of entropy for the process*

H_2O (1, 373 K, 101325 NM^{-2}) → H^2O (v, 373 K, 101325 Nm^{-2})

ΔH_{vap} = 40850 J mol^{-1} at 373 K.

Solution:

The given process is isothermal and isobaric phase transformation.

$$\therefore \Delta S_{vap} = \frac{\Delta H_{vap}}{T_b} = \frac{40850 J\ mol^{-1}}{373K} = \mathbf{109.5 JK^{-1} mol^{-1}}$$

Since ΔS_{vap} = S(vapour) – S (liquid) = 109.5 JK^{-1} mol^{-1}

∴ S(vapour) = S(liquid) + 109.5 JK^{-1} mol^{-1}.

This observation is in tune with the fact that there is more randomness in the vapour state than the liquid state. Hence entropy is a measure of disorderness.

Problem 7:

Evalvate the entropy charges for the following reversible processes:

(a) 1mole H_2O (l, 1atm, 100° C) ⇌ 1 mole H_2O (v, 1 atm, 100°C)

ΔH_{vap} = 40850 J mol^{-1}

(b) 1 mole Sn (a, 1 atm, 13° C) ⇌ 1 mole Sn (β, 1 atm, 13° C)

ΔH_{trans} = 2090 J mol^{-1}

Solution:

$$\text{(a)}\ \Delta S_{vap} = \frac{\Delta H_{vap}}{T_b}$$

Here ΔH_{vap} = 40850 J mol^{-1}

T_b = 100 + 273 = 373 K

$$\therefore \Delta S_{vap} = \frac{40850 Jmol^{-1}}{373K} = 109.5\ JK–1\ mol^{-1}$$

(b) ΔH_{trans} = 2090 J mol^{-1} and T_{trans} = 13 + 273 = 286 K

$$\therefore \Delta S_{trans} = \frac{2090 Jmol^{-1}}{286K} = 7.31\ JK^{-1}mol^{-1}$$

Entropy Change for Reversible Isothermal Expansion of Ideal Gases

Let the volume of n moles of an ideal gas be changed from V_1 to V_2 at constant temperature. Since for this isothermal expansion, the internal energy does not change, the heat absorbed can be calculated from the first law expression as

$$dq_{rev} = -dw_{rev} = PdV = \frac{nRT}{V}dV$$

Therefore, for this process the entropy change of the system can be evaluated as

$$\Delta S = \int_{V_1}^{V_2} \frac{dq_{rev}}{T} = \int_{V_1}^{V_2} nR\left(\frac{dV}{V}\right) = nR\ln\left(\frac{V_2}{V_1}\right) \qquad ...(1)$$

Since at constant temperature $P_1/P_2 = V_2/V_1$, we can also write for an isothermal expansion from pressure P_1 to P_2

$$\Delta S = -nR\ \ln\left(\frac{P_2}{P_1}\right)$$

$$= -nR \times 2.303 \log\left(\frac{P_2}{P_1}\right) \qquad ...(2)$$

Problem 8:

(c) Calculate the entropy change when 1 mole of an ideal gas expands reversibly from an initial volume of 1 dm^3 to a find volume of 10 dm^3 at a constant temperature of 298 K.

Solution:

$$\Delta S = nR\ \ln\left(\frac{V_2}{V_1}\right) = 2.303\ nR \log\left(\frac{V_2}{V_1}\right)$$

Here n = 1; $V_1 = 1\ dm^3$, $V_2 = 10\ dm^3$

$\Delta S = R \ln 10 = 2.303\ R \log 10$

$= 2.303 \times (8.314\ JK^{-1}\ mol^{-1})$

$19.15\ JK^{-1}\ mol^{-1}$.

Entropy Change on Heating or Cooling of a Substance

Let us consider n moles of a substance at a temperature T_1. Let its temperature be change to T_2 keeping the volume constant. If this change is brought about reversibly then we can use the defining equation of entropy. Heat change for this process is

$$dq_{rev} = C_V dT$$

where C_V is the heat capacity at constant volume.

Entropy change is

$$\Delta S = \int_{T_1}^{T_2} \frac{dq_{rev}}{T} = \int_{T_1}^{T_2} \frac{C_V dT}{T}$$

If C_V does not vary with temperature, then

$$\Delta S = C_V \ In \left(\frac{T_2}{T_1}\right) = nC_{V'm} \ In \left(\frac{T_2}{T_1}\right) \quad ...(1)$$

For temperature changes from T_1 to T_2 at constant P,

$$\Delta S = \int_{T_1}^{T_2} C_P \left(\frac{dT}{T}\right) = nC_{P,\,m} \ In \left(\frac{T_2}{T_1}\right) \quad ...(2)$$

if the heat capacity at con 0

stant pressure (C_P) is taken as independent of temperature. Here $nC_{P,m} = C_P$.

Reversible Adiabatic Changes

If any system undergoes a reversible adiabatic change, then by definition, $dq_{rev} = 0$ at all stages of the process, so that

$$\Delta S = q_{rev}/T = 0$$

Therefore, reversible adiabatic processes are often called *isentropic* processes. However, in the case of irreversible adiabatic expansion though qirr is zero but ΔS is not zero, because $\Delta S \neq q_{irr}/T$. The entropy change in a system of an ideal gas being heated or cooled from T_1 to T_2 and expanding from V_1 to V_2, is given by

$$\Delta S = C_V \ In \ \frac{T_2}{T_1} + nR \ In \ \frac{V_2}{V_1}$$

This equation is valid for all types of changes which an ideal gas is permitted to undergo. For adiabatic reversible change.

$$\left(\frac{T_2}{T_1}\right) = \left(\frac{V_1}{V_2}\right)^{nR/CV} \quad \text{for an ideal gas}$$

$$\therefore \ \text{In}\frac{T_2}{T_1} = \frac{nR}{C_V}\text{In}\left(\frac{V_1}{V_2}\right) = \frac{nR}{C_V}\text{In}\left(\frac{V_2}{V_1}\right)$$

$$\text{Hence } \Delta S = -\frac{nR}{C_V}.C_V\text{In}\frac{V_2}{V_1} + nR.\text{In}\frac{V_2}{V_1} = 0$$

That is, any increase in entropy due to expansion is exactly balanced by an equal decrease in entropy due to lowered temperature in an adiabatic reversible expansion of an ideal gas.

$$\frac{T_2}{T_1} = \frac{C_V + (P_2 / P_1)nR}{C_P}$$

Therefore, ΔS for irreversible adiabatic change is not zero, this is so because the final temperature (T_2) in adiabatic irreversible change and adiabatic reversible change are different.

Problem 9:

0.1kg nitrogen gas at 298 K are held by a piston under 30 atm pressure. The pressure is suddenly released to 10 atm and the gas expands adiabatically. If $C_{V,m}$ *= 20.8* J^{K-1} *mol* ^{1}*, what is the final temperature and volume? Calculate ΔS system for this expansion. What would be the value of ΔS (surroundings)?*

Solution:

The change is adiabatic (q = 0) but irreversible as the expansion is sudden against a constant pressure of 10 atm. Applying the first law we shall get

$$C_V (T_2 - /T_1) = w = - P_2 (V_2 - V_1)$$

$$P_1V_1/ T_1 = P_2V_2/T_2 = nR \text{ gives}$$

$$C_V (T_2 - T_1) = nR (T_1P_2/P_1 - T_2)$$

$$\therefore \ T_2 = \frac{nC_{V,m} + nRP_2 / P_1}{nC_{P,m}} \times T_1$$

$$= \left(\frac{20.8JK^{-1}mol^{-1} + 8.314JK^{-1} \times 10/30}{29.114JK^{-1}mol^{-1}}\right) \times 298 \text{ K}$$

= 241 K

and $V_2 = P_1V_1T_2/T_1P_2$

$$= \frac{30 \times V_1 \times 241}{298 \times 10} = \mathbf{2.42\ V_1}$$

Entiopy change for the system is given by

$$\Delta S\text{system} = nC_{V.m}\ \text{In}\frac{T_2}{T_1} + nR\text{In}\frac{V_2}{V_1}$$

$$= n\left(20.8 JK^{-1}mol^{-1} \times 2.303 \log\frac{241K}{298K}\right.$$

$$\left. + 8.314\ JK^{-1}\ mol^{-1}\ 2.303 \times \log\frac{2.42V_1}{V_1}\right)$$

$= n\ (-\ 4.416\ JK^{-1}mol^{-1} + 7.348\ JK^{-1}\ mol^{-1})$

$= 3.571\ mol \times 2.932\ JK^{-1}\ mol^{-1}$

$= 10.47\ JK^{-1}$

$(\because n\ 0.1/0.028 = 3.571\ mol)$

$\Delta S_{Surroundings} = q_{surr}/T = 0$

Problem 10:

Evaluate the change in entropy when 3 moles of a gas are heated from 27° C to 7227° C at a constant pressure of 1 atm. The molar heat capacity of the gas is 23.7 JK^{-1} mol^{-1}

Solution:

$n = 3$ mol. $T_1 = 300$ K, $T_2 = 100$ K,

$C_{P.m} = 23.7\ J_{K-1}\ mol_{-1}$

$\Delta S = nC_{P,m} \times 2.303 \log (T_2/T_1)$

$= (3\ mol) \times (23.7\ JK^{-1}\ mol^{-1}) \times 2.303 \log (100\ K/300\ K)$

$= \mathbf{85.5\ JK^{-1}}$

Problem 11:

Two blocks of copper metal are of the same size but are at different temperatures T_1 and T_2. These block are brought together and allowed to attain thermal equilibrium i.e. both are allowed to come to the same final temperature T. Show that the entropy change is given by

$$\Delta S = C_P \ln\left[\frac{(T_2 - T_1)^2}{4T_1T_2} + 1\right]$$

Solution:

Let the metallic block A be at temperature T_1 and block B at temperature T_2 in the initial state. Let $T_2 > T_1$.

Heat lost by block B = C_{PB} $(T_2 - T)$

Heat gained by block A = C_{PA} $(T_2 - T_1)$

At equilibrium

$$C_{PB} (T_2 - T) = C_{PA} (T - T_1)$$

$$\therefore\ T = \frac{T_1 + T_2}{2} \qquad \because C_{PB} = C_{PA} = C_P$$

Decrease of entropy of block B

$$= \int_{T_2}^{T} C_P d\ln T$$

$$\Delta S_B = C_P \ln\frac{T}{T_2}$$

Increase of entropy of block A

$$\Delta S_A = C_P \ln\frac{T}{T_1}$$

Total entropy change $\Delta S = \Delta S_A + \Delta S_B$

$$= C_P\left[\ln\frac{T}{T_1} + \ln\frac{T}{T_2}\right]$$

$$= C_P\left[\ln\frac{(T_1 + T_2)^2}{4} - \ln T_1T_2\right]$$

$$= C_P \ln\frac{T_1^2 + T_2^2 + 2T_1T_2}{4T_1T_2}$$

Adding and subtracting $2T_1T_2$ to the ln part of the above equation, we get

$$\Delta S = C_P \ln\frac{T_1^2 + T_2^2 + 2T_1T_2 + 4T_1T_2}{4T_1T_2}$$

$$= C_P \ In \left[\frac{(T_2 - T_1)^2}{4T_1T_2} + 1 \right]$$

Now $(T_1 - T_2)^2$ is always positive and hence the quantity within bracket on right side of the above equation is greater than one. Therefore right side is a positive should be spontaneous.

Problem 12:

The molar entropy of an ideal gas ($C_{P, m}$ = 20.9 JK $^{-1}$mol^{-1}) at 298K is 146.0 JK^{-1} mol^{-1}. Find its value at 500 K.

Solution:

We may assume it as a process of heating at constant pressure. Therefore, $dS = dq_{rev}/T = dH/T$ at constant P

$$dS = C_P \ \frac{dT}{T}$$

On integrating we get

$$\int_{S_1}^{S_2} dS = CP \int_{T_1}^{T_2} d \ In \ T$$

$$\therefore \ S\ (T_2) - S\ (T_1) = C_P \ In \ \frac{T_2}{T_1}$$

For the present case

T_1 = 298 K, t_2 = 500 K,

S (298 K) = 146.0JK^{-1} mol^{-1}

$\therefore$ S (500 K) = (146.0 JK^{-1} mol^{-1}) + (20.9 JK^{-1} mol^{-1})

$$\times \ 2.303 \times \log \frac{500K}{298K}$$

= 146.0 JK^{-1} mol^{-1} + 10.81 JK^{-1} mol^{-1}

= 156.81 JK^{-1} mol^{-1}

Problem 13:

Calculate the entropy change when 10 m^3 of an ideal gas ($C_{P, m}$ = 25R) at 27° C and 1.01 × 10^5 N m^{-2} pressure are heated at constant pressure to 127°C.

Solution:

Given that

$T_1 = 300$ K, $T_2 = 400$ K, $C_{P,\,m} = 2.5$ R,

$P = 1.01 \times 10^5$ Nm^{-2},

Heat absorbed dq = dH = CP. dT = TdS

$$\therefore dS = C_P \frac{dT}{T} = C_P d \ln T$$

$$\int_1^2 dS = \Delta S = C_P \int d \ln T = C_P \ln \frac{T_2}{T_1} = nC_{P,m} \ln \frac{T_2}{T_1}$$

$$\text{Now, } n = \frac{PV}{RT} = \frac{(1.01 \times 10^5 \text{Nm}^{-2}) \times 10 \times 10^{-3} \text{m}^3}{(8.314 \text{JK}^{-1}\text{mol}^{-1}) \times (300\text{K})}$$

$$\therefore \Delta S = (0.405 \text{ mol}) \times (2.5 \times 8.314 \text{ mol}^{-1}) \times 2.303 \log \frac{400}{300}$$

$= 2.422$ JK^{-1}

Problem 14:

1 mole of an ideal gas is allowed to expand isothermally at 27° C until its volume is tripled. Calculate ΔS_{syst} and ΔS_{univ} under the following conditions: (a) The expansion is carried out reversibly, (b) the expansion is a free expansion.

Solution:

(a) Isothermal reversible expansion: entropy change can be calculated by using the defining equation

$dS = dq_{rev}/T$ or $\Delta S = q_{rev}/T$.

In the present case

$q = -w = 2.303 \text{ RT} \log V_2/V_1$

$= 2.303 \times 8.314 \times 300 \log 3V_1/V_1$

$= 2740.6$ J mol^{-1}

$\therefore \Delta S_{syst} = q_{rev}/T = 2740.6/300 = 9.135$ JK^{-1} mol^{-1}

$\Delta S_{surr} = -\Delta S_{syst}$; here $\Delta S_{univ} = 0$

(b) In case of free expansion we can not use the equation $dS = dq_{rev}/T$. However, S being a state function, ΔS can be evaluated by

$\Delta S = 2.303\ R \log V_2/V_1 = 9.135\ JK^{-1}\ mol^{-1}$

For surrounding q = 0, and hence $\Delta S_{surr} = 0$

$\Delta S_{univ} = \Delta S_{syst} = \mathbf{9.135\ JK^{-1}\ mol^{-1}}$

Entropy Change for Simultaneous Heating and Expansion of an Ideal Gas

(a) **Temperature and volume are changed.** Let us consider n moles of an ideal gas initially at temperature T_1 and volume V_1. Let its temperature and volume be changed to a final value of T_2 and V_2 respectively. The entropy change can be calculated by devising the following reversible paths for the process.

Path A

Step 1. Expand the gas isothermally at temperature T_1 from volume V_1 to V_2. In this case internal energy change would be zero and the first law will give

$$dq_{rev} = -dw = PdV - \frac{nRT_1}{V}dV$$

$$\therefore \Delta S_1 = \int_{V_1}^{V_2} \frac{dq_{rev}}{T} = \int_{V_1}^{V_2} \frac{nRT_1}{T_1} = nR \ln\left(\frac{V_2}{V_1}\right)$$

Step 2. Heat the gas from T_1 to T_2 at constant volume V_2 then

$$dq_{rev} = C_V dT$$

$$\therefore \Delta S_2 = \int_{T_1}^{T_2} \frac{dq_{rev}}{T} = C_V \int_{T_1}^{T_2} \frac{dT}{T}$$

$$= C_V \ln\left(\frac{T_2}{T_1}\right) = nC_{V,m}\left(\frac{T_2}{T_1}\right)$$

C_V is assumed to be independent of T

The entropy change for the system is

$$\Delta S_{system} = \Delta S_1 + \Delta S_2$$

$$\Delta S_{system} = nR \ln\left(\frac{V_2}{V_1}\right) + nC_{V,m} \ln\left(\frac{T_2}{T_1}\right) \quad ...(1)$$

Path B

Step 1. Heat the gas at constant Volume V_1 from T_1 to T_2

$$\Delta S_1 = nC_{V,m} \ln \left(\frac{T_2}{T_1}\right)$$

Step 2. Expand the gas isothermally at temperature T_2 from volume V_1 to V_2

$$\Delta S_2 = nR \ln \left(\frac{V_2}{V_1}\right)$$

$$\therefore \Delta S_{system} = \Delta S_1 + \Delta S_2 = nC_{V,m} \ln \left(\frac{T_2}{T_1}\right) + nR \ln \left(\frac{V_2}{V_1}\right) \quad ...(2)$$

Thus we see that the final expression for the entropy change comes out to be the same irrespective of the fact whether we first change the volume and then temperature or the reverse. This suggests that entropy is a state function.

Path C

Step 1. Heat the gas at constant volume from T_1 to an intermediate temperature T.

Step 2. Expand isothermally at T from V_1 to an intermediate volume V′.

Step 3. Heat at constant volume from T′ to T_2.

Step 4. Expand isothermally at T_2 from V′ to V_2.

For step 1: $dq_{rev} = nC_{V,m}\, dT$

$\Delta S_1 = nC_{V,m} \ln (T'/T_1)$

For step 2 : $\Delta S_2 = nR \ln (V'/V_1)$

For step 3 : $\Delta S_3 = nC_{V,m} \ln (T_2/T')$

For step 4 : $\Delta S_4 = nR \ln (V2/V')$

The entropy change for the system is

$$\Delta S_{system} = \Delta S_1 + \Delta S_2 + \Delta S_3 + \Delta S_4$$

$$= nC_{V,m} \ln \left(\frac{T'}{T_1}\right) + nR \ln \left(\frac{V'}{V_1}\right) + nC_{V,m} \ln \left(\frac{T_2}{T'}\right) + nR \ln \left(\frac{V_2}{V'}\right)$$

$$= nC_{V,m} \ln \left(\frac{T_2}{T_1}\right) + nR \ln \left(\frac{V_2}{V_1}\right) \quad ...(3)$$

(b) **Temperature and pressure are changed.** Let n moles of an ideal gas, initially at temperature T_1 and pressure P_1 be brought to a final temperature T_2 and pressure P_2. We can use the following reversible path between two states.

Step 1. Heat the gas at constant pressure P_1 from T1 to T_2, then

$$dq_{rev} = nC_{P,\,m}\, dT$$

Step 2. Expand the gas isothermally at temperature T_2 from P_1 to P_2, then

$$dq_{rev} = -\, dw = PdV$$

For step 1: $\Delta S_1 = \int_{T_1}^{T_2} nC_{P,m} \frac{dT}{T}$

$$= nC_{P,\,m}\ \text{In}\left(\frac{T_2}{T_1}\right)$$

(If C_P is independent of T).

For step 2: $\Delta S_2 = \int_{P_1}^{P_2} \frac{PdV}{T_2} = -\int_{P_1}^{P_2} \frac{VdP}{T_2} = -\int_{P_1}^{P_2} nRd\ \text{In}\ P$

$$= -\ nR\ \text{In}\left(\frac{P_2}{P_1}\right)$$

$$\Delta S = \Delta S_1 + \Delta S_2 = nC_{P,\,m}\ \text{In}\left(\frac{T_2}{T_1}\right) - nR\ \text{In}\left(\frac{P_2}{P_1}\right) \quad ...(4)$$

This result can be transformed into the form of equation (4) by using the following relations:

$$C_{P,\,m} = C_{V,\,m} + R$$

$$\frac{P_2}{P_1} = \frac{T_2 V_1}{T_1 V_2}$$

$$\therefore\ \Delta S = n\ (C_{V,m} + R)\ \text{In}\left(\frac{T_2}{T_1}\right) - nR\ \text{In}\left(\frac{T_2}{T_1}\right) - nR\ \text{In}\left(\frac{V_1}{V_2}\right)$$

$$\Delta S = nC_{V,\,m}\ \text{In}\left(\frac{T_2}{T_1}\right) + nR\ \text{In}\left(\frac{V_2}{V_1}\right)$$

Entropy change when a van der Waals gas is heated and expanded simultaneously. Let n moles van der Waals gas initially at temperature T_1 and volume V_1 be brought to temperature T_2 and volume V_2. Combined first and second laws give,

$$TdS = dq_{rev} = dE + PdV \quad ...(1)$$

But $E = f(T, V)$

$$\therefore \; dE = \left(\frac{\partial E}{\partial T}\right)_V dT + \left(\frac{\partial E}{\partial V}\right)_T dV$$

$$= C_V \, dT + \left(\frac{\partial E}{\partial T}\right)_V dV \quad ...(2)$$

From equation (1) and (2), we get

$$dS = \frac{C_V}{T} dT + \frac{1}{T}\left[\left(\frac{\partial E}{\partial V}\right)_T + P\right] dV \quad ...(3)$$

For a van der Waals gas

$$P = \frac{nRT}{V - nb} - \frac{an^2}{V^2}$$

and $\left(\frac{\partial E}{\partial V}\right)_T = \frac{an^2}{V^2}$

$$\therefore \quad dS = \frac{nC_{V,m}}{T} dT + \frac{nR}{V - nb} dV$$

Taking $C_{V, m}$ as independent of T and integrating between suitable limits, we have

$$S_2 - S_1 = \Delta S = nC_{V, m} \ln \frac{T_2}{T_1} + nR \ln \left[\frac{V_2 - nb}{V_1 - nb}\right] \quad ...(4)$$

If the gas expands isothermally then $dT = 0$ and the entropy change is given by

$$\Delta S = nR \ln \left[\frac{V_2 - nb}{V_1 - nb}\right]$$

ΔS can be calculated if b is known. Here we see that the entropy change depends upon the size of the molecules (b).

If $C_{V, m}$ depends upon temperature as given by

$C_{V, m} = \alpha + \beta T + \gamma T_2$, then

$C_{V,m} / T = \alpha/T + \beta + \gamma T$

and $\int_{T_1}^{T_2} \frac{C_{V,m}}{T} dT = \alpha \int_{T_1}^{T_2} \frac{dT}{T} + \beta \int_{T_1}^{T_2} dT + \gamma \int_{T_1}^{T_2} TdT$

$$= \alpha \ln (T_2/T_1) + \beta (T_2 - T_1) + \frac{1}{2}\gamma\left(T_2^2 - T_1^2\right)$$

Therefore, for the n moles of a van der Waals gas the entropy change is given by

$$\Delta S = n\alpha \ln (T_2/T_1) + n\beta (T_2/T_1) + \frac{1}{2}\gamma\left(T_2^2 - T_1^2\right) + nR \ln \left[\frac{V_2 - nb}{V_1 - nb}\right]$$

Similarly, for a gas obeying the equation of state, $\frac{PV}{RT} = 1 + \frac{B(T)}{V}$, we can calculate ΔS.

We have derived

$$dS = \frac{C_V}{T} dT + \frac{1}{T}\left[\left(\frac{\partial E}{\partial V}\right)_T + P\right] dV$$

and $$\left[\frac{\partial E}{\partial V}\right]_T = T\left[\frac{\partial P}{\partial T}\right]_V - P$$

$$\therefore \qquad dS = \frac{C_V dT}{T} + \left[\frac{\partial P}{\partial T}\right]_V dV \qquad ...(5)$$

For the given gas

$$\left(\frac{\partial P}{\partial T}\right)_V = \frac{R}{V} + \frac{R}{V^2} B(T) + \frac{RT}{V^2}\left(\frac{dB}{dT}\right)_V \qquad ...(6)$$

Equations (6) and (5) give,

$$dS = \frac{C_V}{T} dT + \left[\frac{R}{V} + \frac{R}{V^2} B(T) + \frac{RT}{V^2}\left(\frac{dB}{dT}\right)_V\right] dV \qquad ...(7)$$

Integrating equation (8) we get,

$$\Delta S = C_V \ln \frac{T_2}{T_1} + R \ln \frac{V_2}{V_1} - RB\left[\frac{1}{V_2} - \frac{1}{V_1}\right] - RT\frac{dB}{dT}\left[\frac{1}{V_2} - \frac{1}{V_1}\right]$$

$$\Delta S = C_V \ln \frac{T_2}{T_1} + R \ln \left[\frac{V_2}{V_1}\right] + R\left[\left(B + T\frac{dB}{dT}\right)\right]\left[\frac{1}{V_1} - \frac{1}{V_2}\right]$$

Thus, it is possible to calculate the entropy change for a gas obeying virial equation of state, provided we know the experimental conditions and the data for B (T).

Problem 15:

One mole of an ideal gas is taken through a reversible carnot cycle. Predict the sign of ΔT, ΔP, ΔV, ΔE, ΔH, ΔS in each step without preforming the actual calculations.

Solution:

A reversible Carnot cycle consists of (a) isothermal expansion, (b) adiabatic expansion, (c) isothermal compression, and (d) adiabatic compression.

(a) $\Delta T = 0$ (isothermal); $\Delta V > 0$ (expansion); $\Delta P < 0$ (since $V \propto 1/P$); $\Delta H = 0$, $\Delta E = 0$ since for ideal gas $E = E(T)$ and $H = H(T)$, $\Delta S > q > 0$ for isothermal expansion.

(b) $\Delta V > 0$, (expansion), $\Delta P < 0$, $\Delta E < 0$, $\Delta H < 0$, $\Delta T < 0$ (adiabatic expansion), $\Delta S = 0$ $(q = 0)$

(c) $\Delta T = 0$, $\Delta V < 0$, $\Delta P > 0$, $\Delta E = 0$, $\Delta H = 0$, $\Delta S < 0$.

(d) $\Delta T > 0$, $\Delta V < 0$, $\Delta P > 0$, $\Delta E > 0$, $\Delta H > 0$, $\Delta S < 0$.

Problem 16:

Calculate the entropy change when 0.5 dm^3 *of an ideal gas* ($C_{V,m}$ $= 12.6\ JK^{-1}\ mol^{-1}$) *at 298 K and 1 atm is allowed to expand to double its volume and simultaneous heated to* 373 K.

Solution:

ΔS for the system is the same whether the changes take place in one step or in stages. Thus ΔS system is given by

$$\Delta S = nC_{V,m} \ln \frac{T_2}{T_1} + nR \ln \frac{V_2}{V_1}$$

$C_{V,m} = 12.6\ JK^{-1}\ mol^{-1}$, $T_1 = 298K$,

$T_2 = 373$ K, $V_1 = 0.5\ dm^3$,

$V_2 = 1.0\ dm^3$,

$P_1 = 1$ atm,

$$n = P_1V_1/RT_1 = \frac{[(1atm) \times 0.5\ dm^3]}{(0.082 dm^3 atm K^{-1} mol^{-1}) \times (298K)}$$

$= 0.02$ mol

$\therefore \Delta S = (0.02 \text{ mol}) \times (12.6 \text{ JK}^{-1}\text{mol}^{-1}) \times 2.303 \log \frac{373\text{K}}{298\text{K}}$

$+ (0.02 \text{ mol}) \times (8.314\text{JK}^{-1}\text{mol}^{-1}) \times 2.303 \log \frac{1.0\text{dm}^3}{0.5\text{dm}^3}$

$= \mathbf{0.173\ JK^{-1}}$.

Problem 17:

10 g ice at 0° C are added to20 g water at 90° C in a thermally insulated flask of negligible of negligible heat capacity. The heat of fusion of ice is 5980 J mol[1]. What is the final temperature, ΔS_{sytem} and $\Delta S_{surroundings}$? $C_{P,\,m}(H_2O, l) = 75.42\ JK^{-1}mol^{-1}$.

Solution:

When ice and water are mixed then ice melts and the temperature of the mixture is changed. The change may be considered step by step as follows:

(i) (10/08) mol H_2O (s, 273 K) $\xrightarrow{\text{Melt}}$ (10/18) mol H_2O (1,273K)

$q\ (i) = n.\ \Delta H_{fus} = 5980 \times \frac{10}{18}\ J = 3322.2\ J$

The water at 273 K is heated to a temperature T_2K.

(ii) (10/18) H_2O (l, 273 K) $\xrightarrow{\text{Heat}}$ (10/18) H_2O (l,T_2K)

$q\ (ii) = nC_{P,m}\ (H_2O)\ (T_2 - T_1)$

$= (10/18)\ (75.42\ JK^{-1}\ mol^{-1})\ (T_2 - 273)$

$= 41.9\ (T_2 - 273)$

(iii) (20/18) mol H_2O (l, 363 K) $\xrightarrow{\text{Cool}}$ (20/18) mol H_2O (l, T_2 K)

$q\ (iii) = (20/18)\ mol \times (75.42\ JK^{-1}\ mol^{-1}) \times (T_2 - 363)$

$= 83.4\ (T_2 - 363)$

$\therefore$ 10g H_2O (s, 0°C) + 20g H_2O (l, 90°C) $\rightarrow$ 30g H_2O (l, T_2K)

Total heat change in the system q = q (i) + q (ii) + q (iii)

Since the change is adiabatic, therefore q = 0, and q (i) + q (ii) + q (iii) = 0

$$\frac{10}{18} \times 5980 + \frac{10}{18} \times 75.42 \times (T_2 - 273) + \frac{20}{18} \times 75.42\ (T_2 - 363) = 0$$

Solving for T_2 we get $T_2 = 306$ K.

Entropy changes:

$$\Delta S\ (i) = q_1/273 = \frac{5980 \times 10}{273 \times 18} = 12.169\ JK^{-1}$$

$$\Delta S\ (ii) = n_1 C_{P,m}\ ln\ (T_2/T_1)$$

$$= \frac{10}{18} \times 75.42\ ln\ \frac{306}{273} = -\ 4.782\ JK^{-1}$$

$$\Delta S\ (iii) = n_2 C_{P,m}\ ln\ (T_2/T_1)$$

$$= \frac{20}{18} \times 75.42\ ln\ \frac{306}{363} = -\ 14.317\ JK^{-1}$$

$$\Delta S = \Delta S\ (i) + \Delta S\ (ii) + \Delta S\ (iii) = 2.634\ JK^{-1}$$

$\Delta S_{Surrounding} = 0$, since the process is adiabatic, $q_{surr} = 0$.

Problem 18:

One mole of supercooled water at – 10°C and 1 atm pressure turns into ice at –10°C. Calculate the entropy change in the surroundings and the system and the net entropy change. Heat capacity of water and ice at 1 atm may be taken as constant at 75.42 JK^{-1} mol^{-1} and 37.20 JK^{-1} mol^{-1} respectively.

ΔH_{fus} (273 K) = 6008 J mol^{-1}.

Solution:

The system is H_2O (l, – 10°C, 1 atm) → H_2O (s, –10°C, 1 atm). The given process is not a reversibly one so the entropy change for this cannot be calculated directly. But we can consider a series of reversible steps leading form the liquid water at – 10°C to solid ice at – 10°C.

The various steps are as follows:

Step 1. Heat reversibly the super cooled water at – 10°C to 0°C keeping the pressure constant.

Step 2. Allow the solidification of water at 0°C to ice at 0°C.

Step3. Cool the ice reversibly from 0°C to – 10°C. Thus

Step 1. H_2O (l, –10° C, 1 atm) $\xrightarrow{\text{Heat}}$ H_2O (l, 0° C, 1 atm)

The heat required to raise the temperature of supercooled water by dT is

$dq_{rev} = C_P\, dT = dH$

The entropy change in raising the temperature from 263 K to 273 K

$$\Delta S_1 = \int_{263}^{273} \frac{dq_{rev}}{T} = C_P H_2O(l) \ln \frac{273}{263}$$

$$= 2.303 \times (75.42\ JK^{-1}mol^{-1}) \times \log \frac{273}{263}$$

$$= 2.81\ JK^{-1}\ mol^{-1}$$

Step 2: H_2O (l, 0°C, 1 atm) → H_2O (s, 0°C, 1atm)

$$\Delta S_2 = -\frac{\Delta H_{fus}}{T}$$

$$= -\frac{6008 Jmol^{-1}}{273K} = -22.0\ JK^{-1}mol^{-1}$$

Step 3: H_2O (s, 0°C, 1atm) $\xrightarrow{\text{COOl}}$ H_2O (s, – 10°C, 1atm)

Heat change: $dq_{rev} = C_P (H_2O, s)\, dT = dH$

Entropy change : $\Delta S_2 = C_P (H_2O, s) \ln \frac{263}{2'}$

$$= 2.303 \times 37.20 \times \log \frac{263}{273}$$

$$= -1.38\ JK^{-1}\ mol^{-1}$$

On adding the equations of steps 1, 2 and 3, we get

H_2O (l, –10°C, 1 atm) → H_2O (s, – 10°C, 1 atm)

and $\Delta S = \Delta S_1 + \Delta S_2 + \Delta S_3$

$= 2.81 - 22.0 - 1.38 = -$ **20.57** $JK^{-1}\ mol^{-1}$

Here we see that there has been a decrease in entropy of water upon crystallization at –10°C, though the process is irreversible. This example emphasizes the fact that sign of the entropy change for the system plus surroundings, and not merely for any component, is related to irreversibility. Thus we must consider the entropy change for the system as well as that of surroundings. Now ΔS_{surr} can be calculated by considering the surroundings to be large enough so that the exchange of heat can be considered as a reversible process. Thus

$\Delta S_{surr} = \Delta H_{surr}/T = -\Delta H_{system}/T$

$\Delta H_{syst} = \Delta H_1 + \Delta H_2 + \Delta H_3$

ΔH_1 = Heat gained by the system in heating water from 263 K to 273 K

$= C_P (H_2O, l) \times \Delta T = (74.42 \ JK^{-1} \ mol^{-1}) \times (10 \ K)$

$= 754.2 J \ mol^{-1}$

$\Delta H_2 = -\Delta H, fus = -6008 \ mol^{-1}$

ΔH_3 = Heat lost by the system when it is cooled from 273 K to 263 K

$= C_P (H_2O, s) \times \Delta T = -(37.2 \ JK^{-1} \ mol^{-1}) \times (10K)$

$= -372.0 \ J \ mol^{-1}$

$\Delta H_{syst} = (754.2 - 6008 - 372) \ J \ mol^{-1}$

$= -5626 \ J = -\Delta H_{surr}$

$$\Delta S_{surr} = \frac{5626}{263K} = 21.39 \ JK^{-1}$$

$\Delta S_{universe} = \Delta S_{syst} + \Delta S_{surr} = -20.57 \ JK^{-1} + 21.39 JK^{-1}$

$= +0.82 \ JK^{-1}$

Problem 19:

Illustrate the method to evaluate the entropy change for the following process:

$H_2O(l, 298 \ K, 1.01 \times 10^5 \ Nm^{-2})$

$\rightarrow H_2O \ (v, 398 \ K, 0.5 \times 10^5 \ Nm^{-2})$.

Solution:

The given process is irreversible, therefore, we cannot use the equation $dS = dq_{rev}/T$. However, this system can be split into various reversible systems can be split into various reversible systems for the convenience. This is a valid procedure since S is a state function. We proceed as follows:

(a) Heat reversibly and isobarically the water from 298 K to 373 K, (b) allow the evaporation of water under reversible isothermal and isobaric condition, (c) heat water vapour reversibly and isobarically from 373 K to 398 K, (d) allow the isothermal and reversible expansion of the water vapour. The processes and the corresponding entropy changes are :

(a) H_2O (l, 298 K, 1.01 × 105 Nm^{-2}) $\xrightarrow{\text{Heat}}$ H_2O (l, 373K, 1.01 × 10^5 N m^{-2})

ΔS (a) = $C_{P, m}$ (H_2O, l) In (373/298)

(b) H_2O (l, 373 K, 1.01 × 10^5 Nm^{-2}) $\rightleftharpoons$ H_2O (v, 373 K,

1.01 × 10^5 N m^{-2})

ΔS (b) = ΔH_{vap}/373

(c) H_2O (v, 373 K, 1.01 × 10^5 Nm^{-2}) $\xrightarrow{\text{Heat}}$ H_2O (v, 398 K,

1.01 × 10^5 N m^{-2})

ΔS (c) = $C_{P, m}$ (H_2O, v) In (398/373)

(d) H_2O (v, 398 K, 1.01 × 10^5 Nm^{-3}) $\xrightarrow{\text{Expand}}$ H_2O (v, 398 K,

0.5 × 10^5 N m^{-2})

$$\Delta S(d) = -\ R\ \text{In}\ \frac{0.5\times10^5\ Nm^{-2}}{1.01\times10^5\ Nm^{-2}}$$

Adding equation (a) to (d) we get,

H_2O (l, 298 K, 1.01 × 10^5 Nm^{-2})→H_2O (v, 398 K 0.5×10^5 Nm^{-2})

ΔS = ΔS (a) + ΔS (b) + ΔS (c) + ΔS (d)

$$= C_{P, m}\ (H_2O, l)\ \text{In}\ \frac{373}{298} + \frac{\Delta H_{vap}}{373} + C_{P, m}\ (H_2O, v)$$

$$\text{In}\ \frac{373}{298} + R\ \text{In}\frac{1.01}{0.5}$$

On substituting the values of heat capacities of liquid and water vapour, ΔS can be calculated. In the step (d) we assumed the water vapour to behave as an ideal gas.

Problem 20:

One mole van der Waals gas is allowed to expand isothermally and reversibly at 273 K from an initial volume of 1.0 dm^3 to 50.0 dm^3. Calculated the entropy change for the gas. The van der Waals constants are

a = 6.5 atm dm^6 mol^{-2}

b = 0.056 dm^3 mol^{-1}.

Solution:

Given that the expansion is reversible and isothermal, therefore $\Delta S = q_{rev}/T$ can be used to calculate entropy change.

For 1 mole van der Waals gas

$q_{rev} = RT \ln (V_2 - b)/ (V_1 - b)$

$\therefore \Delta S = R \ln (V_2 - b)/ (V_1 - b)$

$= (8.314\ JK^{-1}\ mol^{-1}) \times 2.303 \times \log \dfrac{(50-0.056)}{(1-0.056)}$

$= 32.97\ JK^{-1} mol^{-1}$

Problem 21:

Two moles of an ideal gas are expanded isothermally at 298 K from a volume of V to 2.5V. Find the value of ΔS_{gas}, and ΔS_{total} the for following:

(a) Reversible expansion.

(b) Irreversible expansion when 400 J mol^{-1} are less absorbed than in (a).

(c) Free expansion.

Solution:

(a) $\Delta S_{gas} = nR \ln V_2/V_1$

$V_2 = 2.5\ V,\ V_1 = V,$

n = 2 mol, hence

$\Delta S_{gas} = 2.303 \times (2\ mol) \times (8.314\ JK^{-1}\ mol^{-1}) \log 2.5$

$= 15.258\ JK^{-1}$

$q_{rev} = nRT \ln V_2/V_1 = 4540.9\ J$

$\Delta S_{surr} = q_{surr}/T = -\ q_{syst}/T$

$= -\ 4540.9\ J/298\ K$

$= -\ 15.238\ JK^{-1}$

$\Delta S_{total} = \Delta S_{gas} + \Delta S_{surr}$

$= 15.238 - 15.238 = 0$

(b) $\Delta S_{gas} = 2.303\ nR \log V_2/V_1 = 15.238\ JK^{-1}$

$q_{irr} = q_{rev} - 2 \times 400 \text{ J mol}^{-1} = 4540.9 - 800$

$= 3740.9 \text{ J}$

$\Delta S_{surr} = q_{surr}/T = - q_{syst}/T$

$= -3740.9 \text{ J}/298 \text{ K}$

$= - 12.533 \text{ JK}^{-1}$

$\Delta S_{total} = \Delta S_{gas} + \Delta S_{surr}$

$= 15.298 - 12.553 = 2.685 \text{ JK}^{-1}$

(c) $\Delta S_{gas} = 2.303 \text{ nR} \log V_2/V_1 = 15.238 \text{ JK}^{-1}$

$q = 0$ hence $\Delta S_{surr} = 0$, and

$\Delta S_{total} = \Delta S_{gas} = 15.238 \text{ JK}^{-1}$.

Problem 22:

Calculate the entropy change for the transformation

12 (s, 1atm, 298 K) $\rightarrow I_{-2}$ (v, 1atm, 457 K)

Given that : ΔH_{fus}, m = 15.68 kJ mol^{-1} at the m.pt 113.6°C

$\Delta H'_{vap, m}$ = 25.522 kJ mol^{-1} at the b.pt 184° C.

$C_{P. m}$ (12, s) = 54.684 + 13.431 × 10–4 (t –25) J deg–1 mol–1

$C_{P, m}$ (12, s) = 81.588 JK^{-1} mol^{-1}.

Solution:

Split the process at

(i) J_2 (s, 1 atm, 298 K) $\xrightarrow{\text{heat}}$ I_2 (s, 1 atm, 113.6°C)

$$\Delta S \text{ (i)} = \int_{298}^{386.6} \frac{C_{P,m(I_2,S)}}{T} dT$$

$C_{P,m}$ (I_2, S) = 54.684 + 13.431 × 10^{-4} (t + 273 – 273 –25)

$$\therefore \Delta S \text{ (i)} = 54.684 \int \frac{dT}{T} + 13.4.31 \times 10^{-4} \int \frac{TdT}{T}$$

$$- 13.431 \times 10^{-4} \times 298 \int \frac{dT}{T}$$

$$= (54.684 - 13.431 \times 10^{-4} \times 298) \text{ In} \frac{386.6}{298}$$

$$+ 13.431 \times 10^{-4} (386.6 - 298)$$

$= 14.248 \ JK^{-1}mol^{-1}$

(ii) I_2 (s, 1atm, 113.6°C) $\xrightarrow{\text{melt}}$ I_2 (l, 1atm, 113.6°C)

ΔS (ii) $= \Delta H_{fus}$, m/T_m = 15648 J mol^{-1}/386.6 K

= 40.474 JK–1 mol–1

(iii) I_2 (l, 1atm, 113.6°C) $\xrightarrow{\text{heat}}$ I_2 (l, 1atm, 184°C)

$$\Delta S \text{ (iii)} = \int C_{P,m}(I_{2,} l)\frac{dT}{t}$$

$= 81.588 \ JK^{-1} \ mol^{-1}$ In (457 K/386.6 K)

$= 13.649 \ JK^{-1} \ mol^{-1}$

(iv) I_2 (l, 1atm, 457 K) $\xrightarrow{\text{vaporise}}$ I_2 (v, 1atm, 457 K)

ΔS (iv) $= \Delta H_{vap,\ m}/T_b$

$= 25522 \ J \ mol^{-1}/457 \ K$

$= 55.846 \ JK^{-1} \ mol^{-1}$

Total entropy change for the transformation

$\Delta S = \Delta S$ (i) + ΔS (ii) + ΔS (iii) + ΔS (iv)

$= 124.218 \ JK^{-1} \ mol^{-1}$.

Problem 23:

If pressure, volume and temperature of one mole of a gas are related as $(P + a/V^2) V = RT$, *show tha* :

(i) *P is a state function*, (ii) *dP is exact differential* (iii) $(\partial P/\partial T)_V (\partial T/\partial V)_P (\partial V/\partial P)_T + 1 = 0$.

Solution:

dP is an exact differential if $\dfrac{\partial^2 P}{\partial V \partial T} = \dfrac{\partial^2 P}{\partial T \partial V}$

$$\text{Now } P = \frac{RT}{V} - \frac{a}{V^2} \quad \text{...(i)}$$

Differentiating equation (i) with respect to V at constant T, we get

$$\left(\frac{\partial P}{\partial V}\right)_T = -\frac{RT}{V^2} + \frac{2a}{V^3} \quad \text{...(ii)}$$

Differentiating equation (ii) with respect to T at constant V, we have,

$$\frac{\partial^2 P}{\partial T \partial V} = -\frac{R}{V^2}$$

First differentiate equation (i) with respect to T at constant V and then differentiate the new equation (iii) with respect to V at constant T, to get

$$\left(\frac{\partial P}{\partial T}\right)_V = \frac{R}{V} \qquad \text{...(iii)}$$

$$\frac{\partial^2 V}{\partial V \partial T} = \frac{R}{V^2}$$

Since $\dfrac{\partial^2 V}{\partial T \partial T} = \dfrac{R}{V^2} = \dfrac{\partial^2 P}{\partial V \partial T}$

therefore, dP is an exact differential.

Problem 24:

If $X = f(T, V)$ and $V = f(T, P)$, then show that

(i) $\left(\frac{\partial X}{\partial P}\right)_T = \left(\frac{\partial X}{\partial V}\right)_T \left(\frac{\partial V}{\partial P}\right)_T$

(ii) $\left(\frac{\partial X}{\partial T}\right)_P = \left(\frac{\partial X}{\partial V}\right)_T \left(\frac{\partial V}{\partial T}\right)_P + \left(\frac{\partial X}{\partial T}\right)_V$

Solution:

If X and V are state functions then the total differential will be given as

$$dX = \left(\frac{\partial X}{\partial V}\right)_V dT + \left(\frac{\partial X}{\partial V}\right)_T dV \qquad \text{...(i)}$$

$$dV = \left(\frac{\partial V}{\partial T}\right)_P dT + \left(\frac{\partial V}{\partial P}\right)_T dP \qquad \text{...(ii)}$$

Replacing dV from eq. (i) by eq. (ii), we get

$$dX = \left(\frac{\partial X}{\partial T}\right)_V dT + \left(\frac{\partial X}{\partial V}\right)_T \left[\left(\frac{\partial V}{\partial T}\right)_P dT + \left(\frac{\partial V}{\partial P}\right)_T dP\right]$$

$$= \left[\left(\frac{\partial X}{\partial T}\right)_V + \left(\frac{\partial X}{\partial V}\right)_T \left(\frac{\partial V}{\partial T}\right)_P\right] dT + \left(\frac{\partial X}{\partial V}\right)_T \left(\frac{\partial V}{\partial P}\right)_T dP \qquad \text{...(iii)}$$

At constant temperature dT = 0 and eq. (iii) reduces to

$$(dx)_T = \left(\frac{\partial X}{\partial V}\right)_T \left(\frac{\partial V}{\partial P}\right)_T (\partial P)_T$$

$$\text{or } \left(\frac{\partial X}{\partial P}\right)_T = \left(\frac{\partial X}{\partial V}\right)_T \left(\frac{\partial V}{\partial P}\right)_T \quad ...(iv)$$

At constant pressure dP = 0 and equation (iii) yields,

$$\left(\frac{\partial X}{\partial T}\right)_P = \left(\frac{\partial X}{\partial T}\right)_V + \left(\frac{\partial X}{\partial V}\right)_T \left(\frac{\partial V}{\partial T}\right)_P \quad ...(v)$$

Cyclic Rule

If P and V are state functions, the expressions for total change in P and V can be used to derive the cyclic rule.

Let P = f (V, T)

$$\therefore dP = \left(\frac{\partial P}{\partial V}\right)_T dV + \left(\frac{\partial P}{\partial T}\right)_V dT \quad ...(4)$$

Let V = f (T, P)

$$\therefore dV = \left(\frac{\partial V}{\partial T}\right)_P dT + \left(\frac{\partial V}{\partial P}\right)_T dP \quad ...(5)$$

Equation (4) for constant pressure (dP = 0) reduces to,

$$\left(\frac{\partial P}{\partial V}\right)_T (\partial V)_P + \left(\frac{\partial P}{\partial T}\right)_V (\partial T)_P = 0$$

Dividing by $(\partial V)P$, we have

$$\left(\frac{\partial P}{\partial V}\right)_T + \left(\frac{\partial P}{\partial T}\right)_V \left(\frac{\partial T}{\partial V}\right)_P = 0$$

Multiplying by $(\partial V/\partial P)_T$ we obtain

$$1 + \left(\frac{\partial P}{\partial T}\right)_V \left(\frac{\partial T}{\partial V}\right)_P \left(\frac{\partial V}{\partial P}\right)_T = 0 \quad ...(6)$$

Equation (5) for constant V (dV = 0) reduces to

$$\left(\frac{\partial V}{\partial T}\right)_P (\partial T)_V + \left(\frac{\partial V}{\partial P}\right)_T (\partial P)_V = 0$$

Dividing by $(\partial P)V$, we have

$$\left(\frac{\partial V}{\partial T}\right)_P \left(\frac{\partial T}{\partial P}\right)_V + \left(\frac{\partial V}{\partial P}\right)_T = 0$$

Multiplying by $(\partial P/\partial V)_T$, we obtain

$$\left(\frac{\partial V}{\partial T}\right)_P \left(\frac{\partial T}{\partial P}\right)_V \left(\frac{\partial P}{\partial V}\right)_T + 1 = 0 \qquad ...(7)$$

Equations (6) and (7) express cyclic rules.

Problem 25:

Calculate a and b for an ideal gas.

Solution:

PV = RT for one mole an ideal gas

Differentiation gives

P dV + V dP = RdT

or $\left(\frac{\partial V}{\partial T}\right)_P = \frac{R}{P}$ [∵ at constant P, dP = 0]

$$\therefore \qquad \alpha = \frac{1}{V}\left(\frac{\partial V}{\partial T}\right)_P = \frac{R}{VP} = \frac{1}{T} = T^{-1}$$

Also $\left(\frac{\partial V}{\partial P}\right)_T = -\frac{V}{P}$ for an ideal gas

$$\therefore \qquad \beta = -\frac{1}{V}\left(\frac{\partial V}{\partial P}\right)_T = \frac{V}{VP} = \frac{1}{P} = P^{-1}$$

Problem 26:

(a) *A function ϕ is defined as $\phi(x, y) = x^2 y^3 + x$. Write its partial derivatives and total differential $d\phi$. Test whether $d\phi$ is an exact differential or not.*

Solution:

$\phi(x, y) = x^2y^3 + x$

Partial derivatives :

$$\left(\frac{\partial \phi}{\partial x}\right)_y = 2xy^3 + 1, \quad \left(\frac{\partial \phi}{\partial y}\right)_x = 3x^2y^2$$

Total differential $d\phi = \left(\frac{\partial \phi}{\partial x}\right)_y dx + \left(\frac{\partial \phi}{\partial y}\right)_x dy$

$= (2xy^3 + 1)\, dx + 3x^2y^2\, dy$

In order that df bean exact differential, we have to prove

$$\frac{\partial^2 \phi}{\partial y \partial x} = \frac{\partial^2 \phi}{\partial x \partial y}$$

Now $\left(\frac{\partial \phi}{\partial x}\right)_y = 2xy^3 + 1$

and $\left(\frac{\partial \phi}{\partial y}\right)_x = 3x^2y^2$ give

$$\frac{\partial}{\partial y}\left[\left(\frac{\partial \phi}{\partial x}\right)_y\right]_x = \frac{\partial}{\partial y}(2xy^3 + 1)_x = 6\, xy^2$$

$$\frac{\partial}{\partial x}\left[\left(\frac{\partial \phi}{\partial y}\right)_x\right]_y = \frac{\partial}{\partial x}\left(3x^2y^2\right)_y = 6xy^2 = \frac{\partial}{\partial y}\left[\left(\frac{\partial \phi}{\partial x}\right)_y\right]_x$$

Hence $d\phi$ is an exact differential.

(b) *Prove that in the equation* $dz = (52\, x^3y + 10\, y^5)\, dx + (13\, x^4 + 50\, xy^4)\, dy$, *dz is an exact differential.*

Solution:

If we write the given equation as $dz = M\, dx + N\, dy$, then

$M = 52\, x^3y + 10\, y^5$; $(\partial M/\partial y)_x = 52\, x^3 + 50\, y^4$;

$N = 13\, x^4 + 50\, xy^4$; $(\partial N/\partial x)_y = 52\, x^3 + 50\, x^3 + 50\, y^4$

Since $(\partial M/\partial y)_x = (N/\partial x)_y$, hence dz is exact differential.

Problem 27:

A gas obeys the equation of state $\left(P + \frac{a}{V^2}\right)V = RT$. *Derive the cyclic rule for this gas.*

Solution:

The given equation of state can be written as

$$P = \frac{RT}{V} - \frac{a}{V^2} \qquad \text{...(i)}$$

Differentiation gives,

$$dP = \frac{R}{V}dT - \frac{RT}{V^2}dV + \frac{2a}{V^3}dV$$

$$= \frac{R}{V}dT - \left(\frac{RT}{V^2} - \frac{2a}{V^3}\right)dV \quad ...(ii)$$

As constant volume dV = 0 and hence equation (ii) yields.

$$\left(\frac{\partial P}{\partial T}\right)_V = \frac{R}{V} \quad ...(iii)$$

When pressure is constant, dP = 0 and equation (ii) reduces to

$$\left(\frac{\partial T}{\partial V}\right) = \left(\frac{RT}{V^2} - \frac{2a}{V^3}\right) \Big/ \frac{R}{V} \quad ...(iv)$$

For constant temperature process, dT = 0 and equation (ii) can be written as

$$\left(\frac{\partial V}{\partial P}\right)_T = -\frac{1}{\left(\frac{RT}{V^2} - \frac{2a}{V^3}\right)} \quad ...(v)$$

On multiplying equations (iii), (iv) and (v) we get

$$\left(\frac{\partial P}{\partial T}\right)_V \left(\frac{\partial T}{\partial V}\right)_P \left(\frac{\partial V}{\partial P}\right)_T$$

$$= -\left(\frac{R}{V}\right) \times \frac{\left(\frac{RT}{V^2} - \frac{2a}{V^3}\right)}{\left(\frac{R}{V}\right)} \times \frac{1}{\left(\frac{RT}{V^2} - \frac{2a}{V^3}\right)} = -1$$

or $\left(\frac{\partial P}{\partial T}\right)_V \left(\frac{\partial T}{\partial V}\right)_P \left(\frac{\partial V}{\partial P}\right)_T + 1 = 0$, This is cyclic rule.

Problem 28:

Derive an expression for the coefficient of thermal expansion (α) *for a gas that obeys van der Waals equation of state.*

Solution:

For one mole of van der Waals gas

$$\left(P + \frac{a}{V^2}\right)(V - b) = RT$$

or $PV^3 - PbV^2 + aV - ab - RTV^2 = 0$

Differentiating with respect to T at constant P, we get,

$$3PV^2\left(\frac{\partial V}{\partial T}\right)_P - 2PbV\left(\frac{\partial V}{\partial P}\right)_P + a\left(\frac{\partial V}{\partial T}\right)_P - 2RTV\left(\frac{\partial V}{\partial T}\right)_P - RV^2 = 0$$

Solving for $\left(\frac{\partial V}{\partial T}\right)_P$, we have

$$\left(\frac{\partial V}{\partial T}\right)_P = \frac{RV^2}{3PV^2 - 2PbV + a - 2RTV} \qquad ...(i)$$

Now α is defined as

$$\alpha = \frac{1}{V}\left(\frac{\partial V}{\partial T}\right)_P \qquad ...(ii)$$

Putting the expression for $(\partial V/\partial T)_P$ from into (ii), we have

$$\alpha = \frac{1}{V}\left[\frac{RV^2}{3PV^2 - 2PbV + a - 2RTV}\right]$$

$$= \frac{R}{3PV - 2Pb + \frac{a}{V} - 2RT} \qquad ...(iii)$$

van der Waals equation can also be written as

$$RT = PV - Pb + \frac{a}{V} - \frac{ab}{V^2} \qquad ...(iv)$$

Putting the expression for RT into equation (iii), we get

$$\alpha = \frac{R}{3PV - 2Pb + \frac{a}{V} - 2\left(PV - Pb + \frac{a}{V} - \frac{ab}{V^2}\right)}$$

$$= R\left[PV - \frac{a}{V} + \frac{2ab}{V^2}\right]^{-1}$$

Problem 29:

(a) *Derive an expression for* β *for a van der Waals gas.*

Solution :

$$\beta = \frac{1}{V}\left(\frac{\partial V}{\partial P}\right)_T$$

van der Waals equation is written as,

$PV^3 - PbV^2 + aV - ab - RTV^2 = 0$

Differentiating with respect to P at constant T,

$$3PV^2\left(\frac{\partial V}{\partial P}\right)_T + V^3 - 2PbV\left(\frac{\partial V}{\partial P}\right)_T - bV^2 + a\left(\frac{\partial V}{\partial P}\right)_T - 2RTV\left(\frac{\partial V}{\partial P}\right)_T = 0$$

Solving for $\left(\frac{\partial V}{\partial P}\right)_T$, we get

$$\left(\frac{\partial V}{\partial P}\right)_T = \frac{V^2(b-V)}{3PV^2 - 2PbV + a - 2RTV}$$

Also, for van der Waals gas,

$$RTV = PV^2 + a - PbV - \frac{ab}{V}$$

$$\therefore \quad \left(\frac{\partial V}{\partial P}\right)_T = \frac{V^2(b-V)}{3PV^2 - 2PbV + a - 2\left(PV^2 + a - PbV - \frac{ab}{V}\right)}$$

$$= \frac{b-V}{P - a/V^2 + 2ab/V^3}$$

$$\therefore \quad \beta = -\frac{1}{V}\left(\frac{\partial V}{\partial P}\right)_T = -\frac{b-V}{PV - a/V + 2ab/V^2}$$

(b) *Show that volume is a state function for a gas obeying the equation*:

$P = RTV^{-1} - aV^{-2}$.

Solution:

Volume (V) is a state function if dV is exact differential; dV is an exact differential if $(\partial^2V/\partial P\,\partial T) = (\partial^2V/\partial T\,\partial P)$. Differentiating equation $P = RTV^{-1} - aV^{-2}$, we get

$dp = RV^{-1}dT - RTV^{-2}\,dV + 2aV^{-3}dV$

$= RV^{-1}dT + (2aV^{-3} - RTV^{-2})dV$

$\therefore\ (\partial V/\partial P)_T = (2aV^{-3} - RTV^{-2})^{-1} = (aV^{-3} - PV^{-1})^{-1}$ ...(i)

and $(\partial V/\partial T)_P = -RV^{-1}/\ (2aV^{-3} - RTV^{-2}) = -\ R\ [aV^{-2} - P]^{-1}$ ...(ii)

Differentiating equation (i) with respect to T at constant P, we get

$$\frac{\partial^2 V}{\partial T \partial P} = -\ (aV^{-3}\ PV^{-1})^{-2} \times \left[-3aV^{-4}\left(\frac{\partial V}{\partial T}\right)_P + PV^{-2}\left(\frac{\partial V}{\partial T}\right)_P\right]$$

$= V^2\ (aV^{-2} - P)^{-2} \times V^{-2}\ (P - 3aV^{-2}) \times R\ (aV^{-2} - P)^{-1}$

$= R\ (aV^{-2} - P)^{-3}\ (P - 3aV^{-2})$ ...(iii)

Differentiating equation (ii) with respect to P at constant T, we have

$$\frac{\partial^2 V}{\partial P \partial T} = R\ (aV\text{–}2\text{–}P)\text{–}2\ \times \left[-2aV^{-3}(\partial V/\partial P)_T{}^{-1}\right]$$

$= -\ R\ (aV^{-2} - P)^{-2} \times [2aV^{-3}(aV^{-3}PV^{-1})^{-1} + 1]$

$= R\ (P - 3aV^{-2})\ (aV^{-2} - P)^{-3}$ ...(iv)

Thus $[\partial^2 V/\partial T\ \partial P] = [\partial^2 V/\partial P\ \partial T]$

Problem 30:

Prove that the volume of an ideal gas is a homogeneous function of zeroth degree in pressure and temperature.

Solution:

Volume of a fixed amount of a gas is a function of T and P,

$V = f\ (T, P)$ ...(i)

Applying the Euler's theorem of homogeneity we have

$$T\frac{\partial V}{\partial T} + P\frac{\partial V}{\partial P} = mV \qquad \text{...(ii)}$$

For an ideal gas of known amount, PV = nRT and

$$\left(\frac{\partial V}{\partial T}\right)_P = \frac{nR}{P},\ \left(\frac{\partial V}{\partial P}\right)_T = -\frac{V}{P}$$

Substituting the expressions for these derivatives into equation (ii) we obtain

$$\frac{nRT}{P} - V = mV \text{ or } \frac{nRT}{P} - \frac{nRT}{P} = mV = 0\ (\because V = nRT/P)$$

Since V is not zero, therefore m must be zero. Hence for an ideal gas volume is a homogeneous function of the zeroth degree in pressure and temperature *i.e.*,

$$V(\lambda P, \lambda T) = \lambda^{o} V(T, P)$$

This is obvious from the relation V = nRT/P also. Suppose we double both the temperature and pressure, the volume of the ideal gas will not be changed.

For other extensive properties we shall have

(i) $A = A(T, V, n_1, n_2)$ ...Helmholtz free energy

$\therefore A(T, \lambda V, \lambda n_1, \lambda n_2) = \lambda A$

since T is an intensive property, therefore, it is not multiplied by λ whereas V being an extensive property is multiplied by λ when the moles (n_i's) are so multiplied.

$S = S(E, V, n_1, n_2)$ ...Entropy

$S(\lambda E, \lambda V, \lambda n_1, \lambda n_2) = \lambda S$

$G = G(T, P, n_1, n_2)$

$G(T, P, \lambda n_1, \lambda n_2) = \lambda G$ ...Gibbs free energy

T and Pare intensive variables and hence are not multiplied by λ.

(iv) $H = H(S, P, n_1, n_2)$ ...Enthalpy

$\therefore H(\lambda S, P, \lambda n_1, \lambda n_2) = \lambda H$

(v) $E = E(S, V, n_1, n_2)$ ...Intrinsic energy

$\therefore E(\lambda S, \lambda V, \lambda n_1, \lambda n_2) = \lambda E$

(vi) $V = V(T, P, n_1, n_2)$ ...Volume

$\therefore V(T, P, \lambda n_1, \lambda n_2) = \lambda V$

The treatment of homogeneity enables us to define the intensive and extensive properties in more precise form. Let us consider a consider a set of intensive variables x, y, z, and a set of extensive variables X, Y, Z, then

(i) The function F (x, y, z,...X, Y, Z,..) is an extensive variable if and only if $F(x, y, z, \lambda X, \lambda Y, \lambda Z) = \lambda F(x, y, z, X, Y, Z)$ for arbitrary positive λ.

For example, volume is an extensive property because

$$V(T, P, \lambda n_1, \lambda n_2) = \lambda V(T, P, n_1, n_2).$$

(ii) The function f (x, y, z, X, Y,Z) is an intensive property if and only if f (x, y, z, λX, λY, λZ) = f (x, y, z, X, Y, Z) for arbitrary positive λ.

Problem 31:

Prove that 1/T is an integrating factor for

$$dq = \frac{3}{2}nRdT + \frac{nRT}{V}dV$$

for an ideal gas.

Solution:

$$\text{Given that } dq = \frac{3}{2}nRdT + \frac{nRT}{V}dV = M\,dT + N\,dV$$

where $M = \frac{3}{2}nR$ and $N = \frac{nRT}{V}$

$$\text{Now } \left(\frac{\partial M}{\partial V}\right)_T = \frac{\partial}{\partial V}\left(\frac{3}{2}nR\right)_T = 0$$

$$\text{and } \left(\frac{\partial N}{\partial T}\right)_V = \frac{\partial}{\partial T}\left(\frac{nRT}{V}\right) = \frac{nR}{V}$$

It is seen that $\left(\frac{\partial M}{\partial V}\right)_T \neq \left(\frac{\partial N}{\partial T}\right)_V$ in the present case.

Let $f = T^{-1}$ be the integrating factor, then

$$f.\,dq = \frac{3}{2}nR\frac{dT}{T} + \frac{nRT}{V}\left(\frac{1}{T}\right)dV$$

$$= \frac{3}{2}nR\frac{dT}{T} + \frac{nR}{V}dV = M\,dT + NdV$$

$$M = \frac{3}{2}\frac{nR}{T}, N = \frac{nR}{V}$$

$$\therefore \left(\frac{\partial M}{\partial V}\right)_T = \frac{\partial}{\partial V}\left(\frac{3}{2}\frac{nR}{T}\right)_T = 0;$$

$$\text{and} \left(\frac{\partial N}{\partial T}\right)_V = \frac{\partial}{\partial T}\left(\frac{nR}{V}\right)_V = 0$$

$$\text{Thus} \left(\frac{\partial M}{\partial V}\right)_T = \left(\frac{\partial N}{\partial T}\right)_V = 0$$

Therefore, (dq/T) is exact differential and (1/T) is an integrating factcr.

Problem 32:

For an ideal gas (PV = nRT) show that (1/T) is an integrating factor for dw = PdV.

Solution:

Given that dw = PdV ...(i)

PV = nRT for ideal gas.

$\therefore$ PdV + VdP = nRdT

$$\text{and } dV = \frac{nR}{P}dT - \frac{V}{P}dP \qquad \text{...(ii)}$$

Substituting the expression for dV in equation (i) we get

$$dw = nRdT - VdP = nRdT - \frac{nRT}{P}dP$$

$$\text{Now} \quad \frac{\partial}{\partial P}(nR)_T = 0$$

$$\text{and} \quad \frac{\partial}{\partial T}\left(-\frac{nRT}{P}\right)_P = -\frac{nR}{P}$$

Therefore, dw is inexact differential.

Let $f = \frac{1}{T}$ be an integrating factor, then

$$f.\ dw = \frac{1}{T}nR\ dT - \frac{nR}{P}dP$$

$$\therefore \quad \frac{\partial}{\partial P}\left[\frac{nR}{T}\right]_T = 0$$

$$\text{and} \quad \frac{\partial}{\partial T}\left[\frac{nR}{P}\right]_P = 0$$

Thus $f.\ dw = \frac{nR}{T}dT - \frac{nR}{P}dP$ is an exact differential and $\frac{1}{T}$ is and integrating factor.

3

LAWS OF THERMODYNAMICS

INTRODUCTION

Thermodynamics in various fields of science is based on the wide application of the laws of thermodynamics which are as well established, as widely accepted and as free from experimental anomalies as any known scientific laws.

First Law $dE = dq + dw$

The change in internal energy (dE) in any system is the sum of the heat entering the system (dq) and the work done on the system (dw). The internal energy of an isolated system is constant and $dq = 0$, $dw = 0$, $dE = 0$.

Second Law $dS = dq_{rev}/T$

The change in entropy (dS) during any change in a system is equal to the heat entering the system (when the change is performed reversibly) divided by the absolute temperature T. In all actual processes the entropy of an isolated system always increases ($dS > 0$).

Third Law

The entropy at absolute zero (0K) is zero for any pure crystalline substance *i.e.*, S^0 (0K) = 0.

ZEROTH LAW

This law like other laws is based on human experiences. We can explain this law further as follows. Suppose we have three systems A, B and C placed together. A and C are in contact so that the properties of both change and finally the thermal equilibrium is attained and two come at the same temperature; C is in contact with B also then again thermal equilibrium is attained. If this is so then A and B should also be in thermal equilibrium. That is A, B and C should be at the same temperature.

If A (T_A) $\rightleftharpoons$ C (T_C) B (T_B),

then A (T_A) $\rightleftharpoons$ B (T_B), or if $T_A = T_C = T_B$, then $T_A = T_B$.

Thus the law provides a rational definition of temperature.

In thermodynamic studies the temperature's unit is taken as kelvin (K). The kelvin is strictly defined as the fraction 1/273.16 of the temperature interval between the absolute zero (– 273.16°C) and the triple point of water. In practice the temperature in kelvin is obtained by adding 273.15 to the temperature in degree celcius, e.g.,

25°C = 25 + 273. 15 = 298.15 K.

The use of the symbol °K is not recommended. The symbol of temperature in kelvin scale is T.

A thermodynamic system is said to be in thermal equilibrium if any two of its independent thermodynamic co-ordinates X and Y remain constant as long as the external conditions remain unaltered. Consider a gas enclosed in a cylinder fitted with a piston. If the pressure and volume of the enclosed mass of gas are P and V at the temperature of the surroundings, these values of P and V will remain constant as long as the external conditions viz., temperature and pressure remain unaltered. The gas is said to be in thermal equilibrium with the surroundings.

The *zeroth law* of thermodynamics was formulated after the first and the second laws of thermodynamics have been enunciated. This law helps to define the term *temperature* of a system.

This law states that *if, of three systems. A, B and C, A and B are separately in thermal equilibrium with C, then A and B are also in thermal equilibrium with one another.*

Conversely the law can be stated as follows :

If three or more systems are in thermal contact, each to each, by means of diathermal walls and are all in thermal equilibrium together, then any two systems taken separately are in thermal equilibrium with one another.

Consider three fluids A, B and C. Let P_A, F_A represent the pressure and volume of A, P_B, V_B, the pressure and volume of B, and P_C, V_C are the pressure and volume of C.

If A and B are in thermal equilibrium, then

$$\phi_1(P_A, V_A) = \phi_2(P_B, V_B)$$

or $$F_1[P_A, V_A, P_B, V_B] = 0 \quad \text{...(i)}$$

Expression (i) can be solved, and

$$P_B = Af_1[P_A, V_A, V_B] \quad \text{...(ii)}$$

If B and G are in thermal equilibrium

$$\phi_2(P_B, V_B) = \phi_3(P_C, V_C)$$

or $$F_2[P_B, V_B, P_C, V_C] = 0$$

Also $$P_B = f_2[V_B, P_C, V_C] \quad \text{...(iii)}$$

From equations (ii) and (iii) for A and C to be in thermal equilibrium separately,

$$f_1(P_A, V_A, V_B) = f_2[V_B, P_C, V_C) \quad \text{...(iv)}$$

If A and C are in thermal equilibrium with B separately, then according to the zeroth law, A and C are also in thermal equilibrium with one another.

$\therefore$ $$F_3[P_A, V_A, P_C, V_C] = 0. \quad \text{...(v)}$$

Equation (iv) contains a variable V_B, whereas equation (v) does not contain the variable V_B. It means

$$\phi_1(P_A, V_A) = \varphi_3(P_C, V_C) \quad \text{...(vi)}$$

In general,

$$\phi_1(P_A, V_A) = \phi_2(P_B, V_B) = \phi_3(P_C, V_C) \quad \text{...(vii)}$$

These three functions have the same numerical value though the parameters (P, V) of each are different. This numerical value is termed as *temperature* (T) of the body.

$\therefore$ $$\phi(P, V) = T \quad \text{...(viii)}$$

This is called the equation of state of the fluid.

Therefore, the temperature of a system can be defined as the property that determines whether or not the body is in thermal equilibrium with the neighbouring systems. If a number of systems are in thermal equilibrium, this common property of the system can be represented by a single numerical value called the temperature. It means that if two systems are not in thermal equilibrium, they are at different temperatures.

In a mercury in glass thermometer, the pressure above the mercury column is zero and volume of mercury measures the temperature. If a thermometer shows a constant reading in two systems A and B separately, it will show the same reading even when A and B are brought in contact.

FIRST LAW OF THERMODYNAMICS

It is true when the whole of the work done is used in producing heat or vice versa. Here, W = JH where J is the Joule's mechanical equivalent of heat. But in practice, when a certain quantity of heat is supplied to a system the whole of the heat energy may not be converted into work. Part of the heat may be used in doing external work and the rest of the heat might be used in increasing the internal energy of the molecules. Let the quantity of heat supplied to a system be δH, the amount of external work done be δW and the increase in internal energy of the molecules be dU. The term U represents the internal energy of a gas due to molecular agitation as well as due to the forces of intermolecular attraction. Mathematically

$$\delta H = dU + \delta W \qquad ...(i)$$

Equation (i) represents the first law of thermodynamics. All the quantities are measured in heat units. The first law of thermodynamics states that the amount of heat given to a system is equal to the sum of the increase in the internal energy of the system and the external work done.

For a cyclic process, the change in the internal energy of the system is zero because the system is brought back to the original condition. Therefore for a cyclic process $\oint \delta U = 0$

$$\text{and} \qquad \oint \delta H = \oint dw \qquad ...(ii)$$

[Both are expressed in heat units].

This equation represents Joule's law.

For a system carried through a cyclic process, its initial and final internal energies are equal. From the first law of thermodynamics, for a system undergoing any number of complete cycles

$$U_2 - U_1 = 0$$

$$\oint \delta H = \oint \delta W$$

$$H = W \qquad \text{[Both are in heat units]}$$

FIRST LAW OF THERMODYNAMICS FOR A CHANGE IN STATE OF A CLOSED SYSTEM

For a closed system during a complete cycle, the first law of thermodynamics is written as

$$\oint \delta H = \oint \delta W$$

In practice, however, we are also concerned with a process rather than a cycle. Let the system undergo a cycle, changing its state from 1 to 2 along the path A and from 2 to 1 along the path B.

According to the first law of thermodynamics

$$\oint \delta H = \oint \delta W$$

For the complete cyclic process

$$\int_{1A}^{2A} \delta H + \int_{2B}^{1B} \delta H = \int_{1A}^{2A} \delta W + \int_{2B}^{1B} \delta W \qquad \text{...(i)}$$

Now, consider the second cycle in which the system changes from state I to state 2 along the path A and returns from state 2 to state 1 along the path C. For this cyclic process

$$\int_{1A}^{2A} \delta H + \int_{2C}^{1C} \delta H = \int_{1A}^{2A} \delta W + \int_{2C}^{1C} \delta W \qquad \text{...(ii)}$$

Subtracting (ii) from (i)

$$\int_{2B}^{1B} \delta H - \int_{2C}^{1C} \delta H = \int_{2B}^{1B} \delta W - \int_{2C}^{1C} \delta W$$

or

$$\int_{2B}^{1B} (\delta H - \delta W) = \int_{2C}^{1C} (\delta H - \delta W) \qquad \text{...(ii)}$$

Here B and C represent arbitrary processes between the states 1 and 2. Therefore, it can be concluded that the quantity ($\delta H - \delta W$) is the same for all processes between the states 1 and 2. The quantity ($\delta H - \delta W$) depends only on the initial and the final states of the system and is independent of the path followed between the two states.

Let $\qquad dE = (\delta H - \delta W)$

From the above logic, it can be seen that

$$\int_1^2 dE = \text{constant and is independent of the path.}$$

This naturally suggests that E is a point function and dE is an exact differential.

The point function E is a property of the system.

Here dE is the derivative of E and it is an exact differential.

$$\therefore \qquad \delta H - \delta W = dE \qquad ...(iv)$$

$$\text{or} \qquad \delta H = dE + \delta W \qquad ...\{v)$$

Integrating equation (v), from the initial state 1 to the final state 2

$$_1H_2 = (E_2 - E_1) + {}_1W_2$$

Note: $_1H_2$ cannot be written as $(H_2 - H_1)$, because it depends upon the path.

Similarly, $_1W_2$ cannot be written as $(W_2 - W_1)$, because it also depends upon the path.

Here $_1H_2$ represents the heat transferred,

$_1W_2$ represents the work done,

E_2 represents the total energy of the system in state 2,

E_1 represents the total energy of the system in state 1.

At this point, it is worthwhile discussing what this E can possibly mean. With reference to the system, the energies crossing the boundaries are all taken care of in the form of B and W. For dimensional stability of Eq. (v), this E must be energy and this must belong to the system. Therefore,

E_2 represents the energy of the system in state 2

E_1 represents the energy of the system in state 1

This energy E acquires a value at any given equilibrium condition by virtue of its thermodynamic state. The working substance, for example a gas, has molecules moving in all random fashion. The molecules have energy associated by virtue of mutual attraction and this part is similar to the potential energy of a body in macroscopic terms. They also have velocities and hence kinetic energy. This energy E therefore can be visualised as comprising of molecular potential and kinetic energies in addition to macroscopic potential and kinetic energies, The first part, which owes its existence to the thermodynamic nature is often called the internal energy which is completely dependent on the thermodynamic state and the other two depend on mechanical or physical state of the system.

$E = U + KE + PE$ + Others which depend upon chemical nature etc.

For a closed system (non-chemical) the changes in all others except U are insignificant and

$$dE = dU$$

∴ From equation (v)

$$\delta H = dU + \delta W \qquad \text{...(vi)}$$

Here all the quantities are in consistent units

FIRST LAW OF THERMODYNAMICS SPECIFIC HEAT OF A GAS (T AND V INDEPENDENT)

The internal energy of a system is a angle valued function of the state variables viz., pressure, volume, temperature etc. In the case of a gas, any two of the variables P, V, T are sufficient to define completely its state. If V and T are chosen as the independent variables,

$$U = f(V, T) \qquad \text{...(i)}$$

Differentiating equation (i)

$$dU = \left(\frac{\partial U}{\partial T}\right)_V dT + \left(\frac{\partial U}{\partial V}\right)_T dV \qquad \text{...(ii)}$$

If an amount of heat δH is supplied to a thermodynamical system, say an ideal gas and if the volume increases by dV at a constant pressure P, then according to the first law of thermodynamics

$$\delta H = dU + \delta W$$

Here $\qquad \delta W = P \,.\, dV$

∴ $\qquad \delta H = dU + P \,.\, dV$

Substituting the value of W from equation (ii)

$$\delta H = \left(\frac{\partial U}{\partial T}\right)_V dT + \left(\frac{\partial U}{\partial V}\right)_T dV + P.dV \qquad \text{...(iii)}$$

Dividing both sides by dT

$$\frac{\delta H}{dT} = \left(\frac{\partial U}{\partial T}\right)_V + \left(\frac{\partial U}{\partial V}\right)_T \frac{dV}{dT} + \frac{P.dV}{dT}$$

or $$\left(\frac{\delta H}{dT}\right) = \left(\frac{\partial U}{\partial T}\right)_V + \left[P + \left(\frac{\partial U}{\partial V}\right)_T\right]\frac{dV}{dT} \quad ...(iv)$$

If the gas is heated at constant volume,

$$\left(\frac{\delta H}{\partial T}\right)_V = C_V \quad \text{and} \quad \frac{dV}{dT} = 0$$

$$\therefore \quad \left(\frac{\delta H}{\partial T}\right)_V = \left(\frac{\partial U}{\partial T}\right)_V = C_V \quad ...(v)$$

When the gas is heated at constant pressure,

$$\left(\frac{\delta H}{\partial T}\right)_P = C_P$$

From equation (iv),

$$C_P = \left(\frac{\partial U}{\partial T}\right)_V + \left[P + \left(\frac{\partial U}{\partial V}\right)_T\right]\left(\frac{\partial V}{\partial T}\right)_P$$

$$C_P = C_V + \left[P + \left(\frac{\partial U}{\partial V}\right)_T\right]\left(\frac{\partial V}{\partial T}\right)_P$$

or $$C_P - C_V = \left[P + \left(\frac{\partial U}{\partial V}\right)_T\right]\left(\frac{\partial V}{\partial T}\right)_P \quad ...(vi)$$

From Joule's experiment, for an ideal gas on opening the stopcock, no work was done and no heat transfer tool; place.

So $\delta H = 0 = \delta U + 0$. Therefore, $dU = 0$. Even though the volume changed while the temperature is constant, there if no change in internal energy.

$$\left(\frac{\partial V}{\partial T}\right)_P = 0$$

From the ideal gas equation

$$PV = RT$$

or $$P\left(\frac{\partial V}{\partial T}\right)_P = R \quad ...(vii)$$

$$\therefore \quad C_P - C_V = P\left(\frac{\partial V}{\partial T}\right)_P + \left(\frac{\partial U}{\partial V}\right)_T\left(\frac{\partial V}{\partial T}\right)_P$$

But $$\left(\frac{\partial U}{\partial V}\right)_T = 0$$

$$\therefore \quad C_P - C_V = P\left(\frac{\partial V}{\partial T}\right)_P = R$$

$$\therefore \quad C_P - C_V = R \qquad \text{...(viii)}$$

Here C_P, C_V and R are expressed in the same units.

From equation (iii)

$$\delta H = \left(\frac{\partial V}{\partial T}\right)_V dT + \left[P + \left(\frac{\partial U}{\partial V}\right)_T\right] dV \qquad \text{...(ix)}$$

For a process at constant temperature

$$dT = 0$$

$$\therefore \quad (\delta H)_T = P(dV)_T + \left(\frac{\partial U}{\partial V}\right)_T (dV)_T \qquad \text{...(x)}$$

This equation represents the amount of heat energy supplied to a system in an isothermal reversible process and is equal to the sum of the work done by the system and the increase in its internal energy.

For a reversible adiabatic process

$$\delta H = 0,$$

Therefore, from equation (ix),

$$0 = C_V\, dT + \left[P + \left(\frac{\partial U}{\partial V}\right)_T\right] dV$$

or

$$C_V dT = -\left[P + \left(\frac{\partial U}{\partial V}\right)_T\right] dV$$

Dividing throughout by dV,

$$C_V\left(\frac{\partial T}{\partial V}\right) = -\left[P + \left(\frac{\partial U}{\partial V}\right)_T\right] \qquad \text{...(xii)}$$

The isobaric volume coefficient of expansion

$$\alpha = \frac{1}{V}\left(\frac{\partial V}{\partial T}\right)_P$$

$$\therefore \quad \alpha V = \left(\frac{\partial V}{\partial T}\right)_P \quad C_P - C_V = P\left(\frac{\partial V}{\partial T}\right)_P$$

$$\therefore \quad \frac{C_P - C_V}{\alpha V} = P$$

But

$$\left(\frac{\partial U}{\partial V}\right)_T = 0 = P - P$$

or $$\left(\frac{\partial U}{\partial V}\right)_T = \left(\frac{C_P - C_V}{\alpha V}\right) - P \qquad \text{...(xiii)}$$

or $$-\left(\frac{C_P - C_V}{\alpha V}\right) = \left[P + \left(\frac{\partial U}{\partial V}\right)_T\right] \qquad \text{...(xiv)}$$

From equations (xii) and (xiv)

$$C_V\left(\frac{\partial T}{\partial V}\right) = -\left(\frac{C_P - C_V}{\alpha V}\right)$$

or $$\left(\frac{\partial T}{\partial V}\right) = \frac{C_V - C_P}{\alpha V C_V} \qquad \text{...(xv)}$$

This expression holds good for an adiabatic reversible process.

RELEVANCE OF THE SECOND LAW

The original statement of the second law of thermodynamics is much concerned with the development of the theoretical aspects of heat engines specially the Carnot's cycle. This has got a very little implication in chemical processes. But, the further developments and various derivations involving the concept of this law are of immense use for chemists. For example, the direction of a particular spontaneous transformation can be ascertained with the help of second law of thermodynamics which is not predicted by first law of thermodynamics.

Spontaneous Process

Most of the natural processes are spontaneous. In fact all human activities and all that happens in the universe involve a significant degree of this concept. For example, the production of wheat from seeds, the formation of clouds in the sky on a sunny day, the fall of rain from clouds, the growth and multiplication of micro-organism when food is digested, are all spontaneous processes.

All those processes which take place of their own accord are called spontaneous process. As such all the spontaneous processes are irreversible in nature. Once having undergone a spontaneous transformation in a particular direction, the process will not take place in the reverse direction spontaneously. The spontaneity however never means that the process is instantaneous. It simply stands that the change is favourable in a particular direction. A few other examples of spontaneous processes are as follows. These will explain as to why do we need a *second law*? A plate dropped on to the floor breaks and the pieces scatter; when we take ice-cream our lips get cold; petroleum vapour and oxygen when sparked react in a car engine producing exhaust gases; water runs downhill; a drop of perfume in a room diffuses quickly to effect the entire atmosphere; aqueous solutions of NaOH and HCl react to produce NaCl (aq) and heat. All these processes are known to proceed in one direction. The question is that why the reverse of the considered processes do not occur? Can we believe on our eyes to see the pieces of broken plate spontaneously reorganise themselves to reproduce the original plate? Has any one experienced ever hot while taking ice cream? Is it possible exhaust gases CO, CO_2, H_2O can recombine to produce petroleum and oxygen? Is it possible for the dispersed vapours of the perfume to get converted into the droplets and return to the bottle under ordinary conditions? Can water flow of itself upwards? Can we get aqueous solutions of NaOH and HCl on dissolving NaCl in water?

None of these processes is known to be spontaneous as we know by our experiences. However, in the light of the first law of thermodynamics none of these processes seem to be impossible, because in these processes the energy in conserved, converted from one form into another. This implies that the first law of thermodynamics is incapable of predicting the direction of a process. Now the question arises that what law of nature compels the spontaneous processes to proceed in a particular direction? How and why the systems are aware of the directions in which to proceed? What is the driving force that leads the process in a given direction? The answers to all these questions can be provided by considering the second law of thermodynamics.

SOME POINTS OF THE SECOND LAW OF THERMODYNAMICS

1. "Heat cannot spontaneously pass from a colder to a warmer body." (R.J.E. Clausius)
2. "It is impossible to transfer heat from a colder system to a warmer without other simultaneous changes occurring in the two systems, or in their environment." (P.S. Epstein)
3. "Every system which is left to itself will, on the average, change toward a condition of maximum probability." (G. N. Lewis)
4. "The state of maximum entropy is the most stable state for an isolated system." (Enrico Fermi).

5. "In any irreversible process the total entropy of all bodies concerned is increased." (G. N. Lewis)
6. "In an adiabatic process the entropy either increases or remains unchanged:

$$\Delta S \geq 0$$

Where the upper sign (>) refers to the irreversible, the lower (=) to the reversible case." (P. S. Epstein)
7. "The entropy function of a system of bodies tends of increase in all physical and chemical processes occurring in nature, if we include in the system all such bodies which are affected by the change." (Saha)
8. "The entropy increases towards a maximum and the energy of the universe in constant." (R.J.E. Clausius)
9. "Gain in information is loss in entropy." (G. N. Lewis)
10. "Entropy is times arrow." (A. Eddington)
11. "Every system left to itself changes rapidly or slowly in such a way to approach a definite final state of rest. No system of its own will change away from the state of equilibrium, except through the influence of external agencies" (G. N. Lewis)
12. " There is a general tendency in nature for energy to pass from more available to less available forms." (J. A. V. Butter)
13. "When an actual process occurs it is impossible to invent a means of restoring every system concerned to its original condition." (G. N. Lewis)
14. "There exists a characteristic thermodynamic function called entropy. The difference of a system in states (1) and (2) is given by the expression

$$S_2 - S_1 = \int_1^2 dq/T$$

over any reversible path,

and $S_2 - S_1 > \int_1^2 dq/T$

over any irreversible path.

The entropy is a property of the state only. Its value for an isolated system can never decrease." (R. E. Gibson)

COMBINED FROM OF THE FIRST AND SECOND LAWS OF THERMODYNAMICS

The mathematical form of the first law is

$$dq = dE - dw = dE + PdV \qquad ...(1)$$

(if only PV work is considered).

The mathematical expression for second law is

$$dq_{rev} = TdS \qquad ...(2)$$

Equation (1) and (2) give

$$TdS = dE + PdV$$

$$\text{or } dE = TdS - PdV \qquad ...(3)$$

This is the combined form of the first and the second laws of thermodynamics. This is very important and useful in the deduction of various other concepts of thermodynamic functions.

Enthalpy and Entropy

Enthalpy is expressed as

$$H = E + PV$$

$$\therefore\ dH = dE + PdV + VdP$$

Putting $dE + PdV = TdS$ we get

$$dH = TdS + VdP \qquad ...(4)$$

Energy as a Function of S and V

Equation (3) clearly indicates that the change in energy depends on the change of entropy (dS) and the change of volume (dV). Thus mathematically E is a function of S and V

$$E = f(S, V)$$

The total differential of E is given as

$$dE = \left(\frac{\partial E}{\partial S}\right)_V dS + \left(\frac{\partial E}{\partial V}\right)_S dV \qquad ...(5)$$

$$\text{Also } dE = TdS - PdV \qquad ...(6)$$

Comparing the coefficients of dS and dV separately in equations (5) and (6) we get,

$$\left(\frac{\partial E}{\partial S}\right)_V = T \qquad ...(7)$$

and $$\left(\frac{\partial E}{\partial V}\right)_s = -P \qquad ...(8)$$

Since dE is an exact differential therefore Euler's of exactness when applied with equation (7) gives

$$\left(\frac{\partial T}{\partial V}\right)_s = -\left(\frac{\partial P}{\partial S}\right)_V$$

From equation (3) we see that for constant energy dE = 0 and

$$\left(\frac{\partial S}{\partial V}\right)_E = \frac{P}{T}$$

Enthalpy as a Function of S and P

$$dH = TdS + VdP \qquad ...(9)$$

This equation clearly shows that the change in enthalpy (dH) depends on the changes in entropy (dS) and pressure (dP). Thus

$$H = f(S, P)$$

$$dH = \left(\frac{\partial H}{\partial S}\right)_P dS + \left(\frac{\partial H}{\partial P}\right)_s dP \qquad ...(10)$$

Comparing the coefficients of dS and dP respectively in equations (10) and (9), we obtain

$$\left(\frac{\partial H}{\partial S}\right)_P = T$$

and $$\left(\frac{\partial H}{\partial S}\right)_s = V \qquad ...(11)$$

THERMODYNAMIC RELATIONS BASED ON SECOND LAW

Thermodynamic Equation of State

Any relationship between the thermodynamic function (E or H) and the PVT data is called the thermodynamic equation of state.

(i) **Energy as a function of T and V.** For any reversible process involving only P.V work, the combined form of the first and second laws can be written as

$$TdS = dq_{rev} = dE + PdV \quad ...(1)$$

E is a state function and dE is exact differential, given as

$$dE = \left(\frac{\partial E}{\partial T}\right)_V dT + \left(\frac{\partial E}{\partial V}\right)_T dV = C_V dT + \left(\frac{\partial E}{\partial T}\right)_V dT + \left(\frac{\partial E}{\partial V}\right)_T dV$$

$C_V = (\partial E/\partial T)_V$ = heat capacity at constant volume

$$\therefore\ dS = \frac{C_V}{T}dT + \frac{1}{T}\left[P + \left(\frac{\partial E}{\partial V}\right)_T\right]dV \quad ...(2)$$

Entropy is also a state function S = S (T, V) and dS is exact differential given as,

$$dS = \left(\frac{\partial S}{\partial V}\right)_V dT + \left(\frac{\partial S}{\partial V}\right)_T dV \quad ...(3)$$

Comparing the coefficient of dT and dV separately of equations (2) and (3), we get

$$\left(\frac{\partial S}{\partial T}\right)_V = \frac{C_V}{T} = \frac{1}{T}\left(\frac{\partial E}{\partial T}\right)_V \quad ...(4)$$

$$\left(\frac{\partial S}{\partial V}\right)_V = \frac{1}{T}\left[P + \left(\frac{\partial E}{\partial V}\right)_T\right] \quad ...(5)$$

Differentiating equation (4) with respect to V at constant T, we get

$$\left(\frac{\partial^2 S}{\partial V \partial T}\right) = \frac{\partial}{\partial V}\left(\frac{C_V}{T}\right)_T = \frac{1}{T}\frac{\partial}{\partial V}\left[\left(\frac{\partial E}{\partial T}\right)_V\right]_T \quad ...(6)$$

On differentiating equation (5) with respect to T at constant volume, we get

$$\frac{\partial^2 S}{\partial T \partial V} = \frac{\partial}{\partial T}\left[\frac{1}{T}\left\{P + \left(\frac{\partial E}{\partial V}\right)_T\right\}\right]_V$$

$$= -\frac{1}{T^2}\left(\frac{\partial E}{\partial V}\right)_T + \frac{1}{T}\left(\frac{\partial P}{\partial T}\right)_V + \frac{1}{T}\frac{\partial^2 E}{\partial T \partial V} \quad ...(7)$$

Since S is a state function and dS is exact differential, therefore, Euler's theorem of exactness must be valid i.e.,

$$\frac{\partial^2 S}{\partial V \partial T} = \frac{\partial^2 S}{\partial T \partial V}$$

On applying this condition, equations (7) and (6) give

$$\frac{1}{T}\frac{\partial^2 E}{\partial V \partial T} = -\frac{P}{T^2} - \frac{1}{T^2}\left(\frac{\partial E}{\partial V}\right)_T + \frac{1}{T}\left(\frac{\partial P}{\partial T}\right) + \frac{1}{T}\frac{\partial^2 E}{\partial T \partial V} \qquad ...(8)$$

Since dE is exact differential

$$\therefore \quad \frac{\partial^2 E}{\partial V \partial T} = \frac{\partial^2 E}{\partial T \partial V}$$

Applying this equality in equation (8), we get

$$\left(\frac{\partial E}{\partial V}\right)_T = T\left(\frac{\partial P}{\partial T}\right)_V - P \qquad ...(9)$$

This expression is known as the thermodynamic equation of state and is true for all substances under all conditions and follows directly from the acceptance of the validity of the first and second laws. Equation (9) can be written in another form if we realise that $(\partial P/\partial T)_V = \alpha/\beta$, where $\alpha = V^{-1}(\partial V/\partial T)_P$, and $\beta = -V^{-1}(\partial V/\partial P)_T$.

$$\left(\frac{\partial E}{\partial V}\right)_T = T\frac{\alpha}{\beta} - P$$

Let us apply it to an ideal gas and then to a van der Waals gas.

THERMODYNAMIC EQUATION OF STATE APPLIED TO AN IDEAL GAS

We have argued that the energy of an ideal gas is independent of volume and have defined an ideal gas as one obeying the two relationships $PV = nRT$ and $(\partial E/\partial V)_T = 0$. With the assistance of the thermodynamic equation of state we can now prove the latter condition.

For an ideal gas $PV = nRT$

$$\therefore \quad \left(\frac{\partial P}{\partial T}\right)_V = \frac{nR}{V} = \frac{P}{T}$$

On substituting the expression $(\partial P/\partial T)V = PT$ into equation (9) we get,

$$\left(\frac{\partial E}{\partial V}\right)_T = T\left(\frac{\partial P}{\partial T}\right)_V - P = T\frac{P}{T} - P = 0$$

Van der Waals equation. It is given as

$$P = \frac{nRT}{V-nb} - \frac{an^2}{V^2}$$

where a is the attraction coefficient and b is the excluded volume per mole; a measure of bulk of the gas molecules. Using the thermodynamic equation of state we can deduce an expression for the internal pressure $(\partial E/\partial V)_T$ for the gas.

For van der Waals equation

$$\left(\frac{\partial P}{\partial T}\right)_V = \frac{nR}{V-nb}$$

On substituting this expression into thermodynamic equation of state we obtain

$$\left(\frac{\partial E}{\partial V}\right)_T = \frac{nRT}{V-nb} - P = \frac{nRT}{V-nb}\left(\frac{nRT}{V-nb} - \frac{an^2}{V^2}\right)$$

$$\left(\frac{\partial E}{\partial V}\right)_T = \frac{an^2}{V^2} \qquad ...(10)$$

For one mole of a gas, equation (10) reduces to

$$\left(\frac{\partial E}{\partial V}\right)_T = \frac{a}{V^2}$$

or $\quad dE = \frac{a}{V^2} dV$

Integration gives $\Delta E = -a\left(\frac{1}{V^2} - \frac{1}{V_1}\right)$

The energy of a van der Waals gas is related to the attraction coefficient a. The energy increase with an isothermal expansion of the gas. When the molecules are close to each other, the volume is smaller, the intermolecular forces reduce the energy of the gas.

(ii) **Enthalpy as function of temperature and pressure.**

The enthalpy is defined as

$H = E + PV$

$\therefore\ dH = dE + PdV + VdP$

The combined first and second laws give dE + PdV = TdS, therefore

$$TdS = dH - VdP \qquad ...(11)$$

H being a state function H = H (T, P) its total differential is given by

$$dH = \left(\frac{\partial H}{\partial T}\right)_P dT + \left(\frac{\partial H}{\partial P}\right)_T dP = C_p dT + \left(\frac{\partial H}{\partial T}\right)_P dP \qquad ...(12)$$

Here $C_P = (\partial H/\partial T)_P$

Putting this expression into equation (11), we get

$$dS = \frac{C_P}{T} dT + \frac{1}{T}\left[\left(\frac{\partial H}{\partial P}\right)_T - V\right] dP \qquad ...(13)$$

Since S = f (T, P)

$$\therefore\ dS = \left(\frac{\partial S}{\partial T}\right)_P dT + \left(\frac{\partial S}{\partial P}\right)_T dP \qquad ...(14)$$

Comparing the coefficients of dT and dP separately in equations (13) and (14), we get

$$\left(\frac{\partial S}{\partial T}\right)_P = \frac{C_P}{T} = \frac{1}{T}\left(\frac{\partial H}{\partial T}\right)_P \qquad ...(15)$$

$$\left(\frac{\partial S}{\partial P}\right)_T = z\frac{1}{T}\left[\left(\frac{\partial H}{\partial P}\right)_T - V\right] \qquad ...(16)$$

On differentiating equation (15) with respect to P at constant T, we get

$$\frac{\partial^2 S}{\partial P \partial T} = \frac{1}{T}\frac{\partial^2 H}{\partial P \partial T} \qquad ...(17)$$

On differentiation w.r. to T at constant P equation (16) yields

$$\frac{\partial^2 S}{\partial T \partial P} = -\frac{1}{T^2}\left[\left(\frac{\partial H}{\partial P}\right)_T - V\right] + \frac{1}{T}\left[\frac{\partial^2 H}{\partial T \partial P} - \left(\frac{\partial V}{\partial T}\right)_P\right] \qquad ...(18)$$

Since S and H and state functions, therefore,

$$\frac{\partial^2 S}{\partial T \partial P} = -\frac{\partial^2 S}{\partial T \partial P}, \text{and } \frac{\partial^2 H}{\partial T \partial P} = \frac{\partial^2 H}{\partial P \partial T}$$

Applying this condition, equations (17) and (18) yield

$$\left(\frac{\partial H}{\partial P}\right)_T = V - T\left(\frac{\partial V}{\partial T}\right)_P \qquad ...(19)$$

The expression (19) is thermodynamic equation of state.

For ideal gas PV = nRT

$$\therefore \quad \left(\frac{\partial V}{\partial T}\right)_P = \frac{nR}{P} = \frac{V}{T}$$

From equation (34), we have

$$\left(\frac{\partial H}{\partial P}\right)_T = V - T\left(\frac{\partial V}{\partial T}\right)_P = V - T\frac{V}{T} = 0$$

Now we shall consider a few more applications of the results of the results of this section to some simple cases:

(a) **Relation between C_P and C_V**

We developed expressions for the difference between the heat capacities at constant pressure and at constant volume, as

$$C_P - C_V = \left[\left(\frac{\partial E}{\partial V}\right)_T + P\right]\left(\frac{\partial V}{\partial T}\right)_P \qquad ...(20)$$

$$\text{But} \quad \left(\frac{\partial E}{\partial V}\right)_T = T\left(\frac{\partial P}{\partial T}\right)_V - P \qquad ...(21)$$

$$\therefore \quad C_P = C_V = T\left(\frac{\partial P}{\partial T}\right)_V\left(\frac{\partial V}{\partial T}\right)_P \qquad ...(22)$$

But $(\partial V/\partial T)_P = \alpha V$, where α is the coefficient of thermal expansion. Therefore, equation (22) may be written as,

$$C_P - C_V = TV\alpha\left(\frac{\partial P}{\partial T}\right)_V \qquad ...(23)$$

V is a function of T and P,

V = f (T, P)

$$\therefore dV = \left(\frac{\partial V}{\partial T}\right)_P dT + \left(\frac{\partial V}{\partial P}\right)_T dP$$

At constant volume, this equation reduces to

$$\left(\frac{\partial V}{\partial T}\right)_P (\partial T)_V + \left(\frac{\partial V}{\partial P}\right)_T (\partial P)_V = 0$$

$$\therefore \quad \left(\frac{\partial P}{\partial T}\right)_V = \frac{(\partial V/\partial T)_P}{(\partial V/\partial P)_T} = \frac{\alpha}{\beta} \qquad ...(24).$$

where β is the compressibility coefficient defined as $b = -V^{-1}(\partial V/\partial P)_T$, and $a = V^{-1}(\partial V/\partial T)_P$

Using equations (23) and (24), We get

$$C_P - C_V = VT\frac{\alpha^2}{\beta}$$

This equation is equally true for solids, liquids and gases.

For water the density is maximum at 3.98° C, and $C_P = C_V$. This can be shown with the help of equation (23) and the definition of density (ρ). The density is defined as mass (m) per unit volume:

$\rho = m/V$,

$$\text{hence} \quad \frac{d\rho}{dT} = \frac{d}{dT}(m/V) = -\frac{m}{V^2}(dV/dT)$$

For maximum, the derivative dρ/dT must be zero, so should be its right side. Since m and V^2 are not zero, therefore dV/dT should be zero. Equation (24) gives $C_P - C_V = 0$, and $C_P = C_V$.

For an ideal gas $\alpha = T^{-1}$, $b = P^{-1}$, therefore $C_P - PV/T = nR$.

(b) **Joule-Thomson coefficient for a van der Waals gas**

Joule-Thomson coefficient is the pressure dependence of temperature under isoenthalpic conditions, and is given as

$$\mu\pi = \left(\frac{\partial T}{\partial P}\right)_H = -\frac{1}{C_P}\left(\frac{\partial H}{\partial P}\right)_T \qquad ...(25)$$

From equations (23) and (25), we get,

$$\mu\pi = \frac{1}{C_P}\left[T\left(\frac{\partial V}{\partial T}\right)_P - V\right] = \frac{V}{C_P}(\alpha T - 1)$$

where $\alpha V = (\partial V/\partial T)_P$

The temperature, at which μπ is zero, is called the inversion temperature given by

$T\pi = V(\partial V/\partial T)_P = 1/\alpha$

For one mole, van der Waal's equation of state is written as

$$PV = RT + Pb - \frac{a(V-b)}{V^2}$$

If $V - b = V$ and $V \cong \frac{RT}{P}$ then

$$PV = RT + Pb - \frac{Pa}{RT}$$

and $$V = \frac{RT}{P} + b - \frac{a}{RT}$$

$$\therefore \left(\frac{\partial V}{\partial T}\right)_P = \frac{R}{P} + \frac{a}{RT^2}$$

On substituting the expression for $(\partial V/\partial T)_P$ into (23) we get,

$$\left(\frac{\partial H}{\partial P}\right)_T = V - T\left(\frac{\partial V}{\partial T}\right)_P = V - \frac{RT}{P} - \frac{a}{RT}$$

$$= \left(\frac{RT}{P} + b - \frac{a}{RT}\right) - \frac{RT}{P} - \frac{a}{RT} = b - \frac{2a}{RT}$$

But $(\partial H/\partial P)_T = -C_P\mu\pi$

$$\therefore \mu\pi = \frac{1}{C_P}\left(\frac{2a}{RT} - b\right)$$

For very low temperatures the first term will dominate and the coefficient will be positive, while for higher temperatures the coefficient will be negative. The *inversion temperature* will be $T\pi = \frac{2a}{Rb}$ when mp is zero.

(c) **Variation of C_V with Volume at Constant T**

$$C_V = \left(\frac{\partial E}{\partial T}\right)_V$$

Differentiating with respect to V at constant T, we get

$$\left(\frac{\partial C_V}{\partial V}\right)_T = \frac{\partial}{\partial V}\left[\left(\frac{\partial E}{\partial T}\right)_V\right] = \frac{\partial}{\partial T}\left[\left(\frac{\partial E}{\partial V}\right)_T\right]_V$$

But $\left(\frac{\partial E}{\partial V}\right)_T = T\left(\frac{\partial P}{\partial T}\right)_V - P$

$$\therefore \quad \left(\frac{\partial C_V}{\partial V}\right)_T = \frac{\partial}{\partial T}\left[T\left(\frac{\partial P}{\partial T}\right)_V - P\right]_V$$

$$= T\frac{\partial^2 P}{\partial T^2} + \left(\frac{\partial P}{\partial T}\right)_V - \left(\frac{\partial P}{\partial T}\right)_V = T\left(\frac{\partial^2 P}{\partial T^2}\right)_V$$

For an ideal gas PV = nRT and $(\partial P/\partial T)_V = nR/V$, therefore

$\left(\frac{\partial^2 P}{\partial T^2}\right)_V = 0$, and hence

$\left(\frac{\partial C_V}{\partial V}\right)_T = 0$ for an ideal gas.

For a van der Waals gas $P = \frac{nRT}{V - nb} - \frac{an^2}{V^2}$, and $(\partial P/\partial T)_V = \frac{nR}{V - nb}$, therefore

$\left(\frac{\partial^2 P}{\partial T^2}\right)_V = 0$, and hence

$\left(\frac{\partial C_V}{\partial V}\right)_T = T\left(\frac{\partial^2 P}{\partial T^2}\right)_V = 0$ for a van der Waal's gas.

(d) For *free expansion* of a gas the Joule coefficient μ_J is defined by $\mu_J = (\partial T/\partial V)_E$. With the help of thermodynamic equation of state we can correlate μ_J with some measurable physical properties of a gas. E is a state function and hence

$$dE = \left(\frac{\partial E}{\partial T}\right)_V dT + \left(\frac{\partial E}{\partial V}\right)_T dV = C_V dT + \left(\frac{\partial E}{\partial V}\right)_T dV$$

For constant energy process dE = 0, and

$$C_V (\partial T)_E = -\left(\frac{\partial E}{\partial V}\right)_T (\partial V)_E$$

or $\left(\frac{\partial T}{\partial V}\right)_E = -\frac{1}{C_V}\left(\frac{\partial E}{\partial V}\right)_T$

But $(\partial T/\partial V)_E = \mu_J$ and $(\partial E/\partial V)_T = T(\partial P/\partial T)_V - P$

$$\therefore \mu_J = -\frac{1}{C_V}\left[T\left(\frac{\partial P}{\partial T}\right)_V - P\right]$$

Since $(\partial P/\partial T)_V = \alpha/\beta$

$$= \mu_J = \frac{P - \alpha T/\beta}{C_V}$$

For an ideal gas $\alpha/\beta = P/T$, therefore $\mu_J = 0$. That is, the free expansion of an ideal gas is isothermal.

ENTROPY CHANGE FOR ISOLATED SYSTEM

For reversible isolated processes the change of entropy is zero whereas for irreversible isolated processes the entropy change is greater than zero. This is shown as follows.

Let us suppose that small changes are made in the system reversibly and irreversibly. The work done by the system on the surrounding in the reversible process $(-dw_{rev})$ i.e.

$$(-dw_{rev}) > (-dw_{irr}) \qquad ...(1)$$

Since energy (E) of the system is a state function and dE is exact differential and must be the same for a given change whether it is performed reversibly or irreversibly. That is $dE_{rev} = dE_{irr} = dE$.

From the first law we have,

$$dE = dq_{rev} + dw_{rev}$$

and $dE = d_{irr} + dw_{irr}$

Since $(-dw_{rev}) > (-dw_{irr})$, R. H. S is positive

$$\therefore dq_{rev} > dq_{irr}$$

This is, in any small change a system will absorb more heat under reversible conditions than under irreversible conditions.

On dividing both sides of equation (2) by T, we get

$$\frac{dq_{rev}}{T} > \frac{dq_{irr}}{T} \qquad ...(2)$$

But (dqrev/T) = dS (definition from 2nd law)

$$\therefore\ dS > \frac{dq_{irr}}{T}$$

$$\text{or } dS > \frac{dE - dw_{irr}}{T} \qquad ...(3)$$

$$\text{and } dS = \frac{dE - dw_{rev}}{T} \qquad ...(3b)$$

In an isolated system there is no exchange of energy, matter, heat etc., across the boundary, i.e., dq = 0, dw = 0, dE = 0.

Introducing these conditions into equations (3b) and (3a), we get

$(dS)_{rev} = 0$ and $(dS)_{irr} > 0$

or $(\Delta s)_{rev} = 0$ and $(\Delta s)_{irr} > 0$ ...(4)

The entropy changes expressed by equation (15) indicate as if it were not a state function? But it is. The why ΔS for reversible and irreversible isolated system are different? The explanation is like is like this. An isolated system includes the system under investigation and the surroundings in its immediate vicinity. To calculate any property of an isolated system , therefore, we must account for the properties of both the system and the surroundings. The entropy change for the isolated system includes the changes in the system as well as surroundings, and the total entropy change is called the entropy of the universe.

ΔS (universe) = ΔS (system) + ΔS (surroundings)

For the system $DS_{rev}^{syst} = DS_{irr}^{syst}$. It is the DS (surrounding) which is different for reversible and irreversible changes.

ENTROPY CONCEPT

The concept of entropy is vital for thermodynamics. It is a Greek word with no equivalent in common language. It was invented by R.J.E. Clausius. He denoted it by the symbol S. *Entropy* stands for trope (a greek word meaning change), the suffix 'en' is written to identify it with *energy*. We have an intuitive feel for the concept of energy; a capacity to do work. However, there is no such intuitive grasp of entropy. Nevertheless, we are quite familiar with the effects the entropy has been used for. Entropy is a measure of randomness or disorderness in a molecular system. More is randomness larger is the value of entropy. More regular is the arrangement lower is the entropy. Entropy is also a measure of unavailable energy. Sometimes entropy is also recognised as an index of exhaustion. Its significance can be appreciated by considering some general observations.

In a cyclic process heat cannot be completely converted into work. The efficiency of working is never hundred per cent. The magnitude of work of irreversible expansion is less than the reversible expansion. Why is it so ? Why the entire heat absorbed cannot be converted into useful work? Wherelse the rest of the unused heat has gone? Why the system's capacity to do work is reduced in an irreversible process? The answers to these questions can partly be given as follows.

When heat is absorbed by a gas, the kinetic energy of the gas molecules will be increased and they will move faster, and collide with (i) each other, (ii) the container and (iii) the face of the piston, thereby the piston will be pushed. Only the third type of collision will bring about the expansions will not bring effective expansion of the gas.

Thus some collisions and hence a fraction of heat is not utilized for the expansion of the gas. Certainly the work due to expansion is note exactly equivalent to the amount of heat absorbed. Since the heat goes waste only because of the random motion of the gas molecules, we can correlate the randomness with the unavailable work.

The extent of randomness or the unavailable work is indicated by the entropy of the system. Under an ideal situation, if all the gas molecules were striking the piston perpendicularly then certainly whole of the heat absorbed will be converted into work as shown in Fig. 1(a). But this is only possible if there is a regular and only one possible arrangement of the gas molecules. In other situations when the molecules are in random motion, a fraction of heat is converted into such a factor which is a measure of randomness of the molecular system. *So entropy is the measure of randomness of a molecular system.*

This statement is also in line with the observations at the boiling point of a fluid. Under ordinary conditions when a liquid is heated its temperature increases, at the boiling point the temperature remains constant will whole of the liquid is transformed into vapour state with the continuous supply of heat. The question is that why the temperature at the boiling point remains constant? Where is the heat being utilised? The answer can be given if we consider the molecular arrangement in the liquid and vapour state at a given temperature.

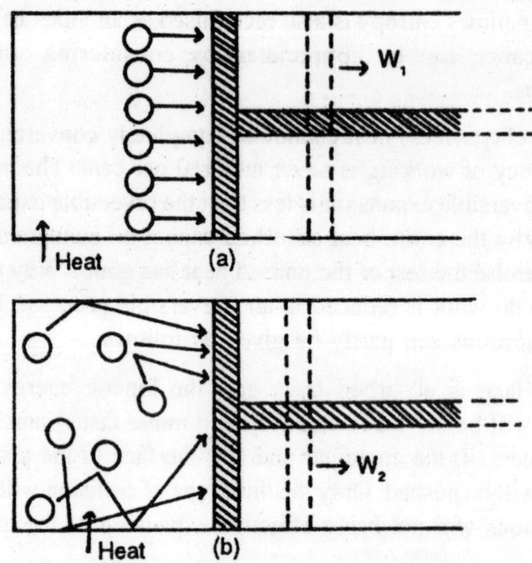

Fig. 1: Conversion of heat work. (a) Ideal case of complete conversion. (b) Partial conversion due to randomness (dotted lines represent the position of the piston after expansion and $w_1 > w_2$).

The molecular arrangement in the liquid state is a bit more regular than the vapour state of the same substance. To disturb the regularity in the liquid state some work must be done. This work done on the system increases the disorderness but not the temperature at the boiling point. Since at the boiling point there is more disorderness in the vapour phase than the liquid, we can say the entropy of the vapour is greater than that of the liquid state of the same substance. Agreeing on the same lines we can say that the entropy of the liquid state is greater than that of the solid state of the same substance. In general, for any substance; S (vapour) > S (liquid) > S (solid) at a given temperature and pressure.

MATHEMATICAL TREATMENT OF ENTROPY CONCEPT

The logical concept of entropy can be given a mathematical foundation. This treatment is essential because we wish to recognise entropy (S) as a thermodynamic function. For this purpose we shall start with the first law of thermodynamics applied to the changes in states of an ideal gas. (Later on this can be extended for other substances also). Let us consider an ideal gas that is undergoing a reversible change from an initial state T_1, V_1 to a final state T_2, V_2. For infinitesimal changes, the first law of thermodynamics can be written as,

$dq_{rev} = dE - dw_{rev}$ ...(i)

If only PV work is considered then

$-dw_{rev} = PdV$

and $dq_{rev} = dE + PdV$

For an ideal gas $dE = C_V\ dT$, and $P = n\ RT/V$, hence

$dq_{rev} = Cv\ dT + (nRT/V)\ dV$ ...(ii)

To know the total heat change we must integrate this equation. But we see that the integration is not possible because the second term on right side of the equation involves two variables T and V, whereas we have to carry the integration with respect to V only. However, if both sides of the equation (ii) are multiplied by l/T, then integration can be carried. Thus

$$\frac{dq_{rev}}{T} = C_V \frac{dT}{T} + nR\frac{dV}{V}$$ ...(iii)

Integrating equation (iii) between the limits

$V = V_1$, $V = V_2$, $T = T_1$, $T = T_2$, we have

$$\int_1^2 \frac{dq_{rev}}{T} = C_V \ln\left(\frac{T_2}{T_1}\right) + nR \ln\left(\frac{V_2}{V_1}\right)$$ ...(iv)

For an ideal gas, $P_1V_1/T_1 = P_2\ V_2/\ T_2$, hence

$$\int_1^2 \frac{dq_{rev}}{T} = C_P \ln\left(\frac{T_2}{T_1}\right) + nR \ln\left(\frac{P_1}{P_2}\right)$$ ...(v)

Here $C_P = (C_V + nR)$ = hear capacity at constant pressure. We find that

(i) dq_{rev} is not an exact differential because

$(\partial\ CV/\partial V)_T \neq (\partial n\ RT/\partial T)v$

(ii) (dq_{rev}/T) becomes an exact differential and can be integrated, hence 1/T is the integrating factor.

(iii) Since $dq_{rev}/\ T$ is an exact differential therefore the function whose change is given by dqrev/T must be a state function. Let this state function be S and hence

$dq_{rev}/T = dS$ ...(vi)

Now the exactness of dS and state dependence of S can be shown as follows:

dS is exact differential. From equations (iii) and (vi) we have

$$dS = (C_V/T)\ dT + (nR/V)dV \qquad ...(viii)$$

At constant T, dT = 0 and

$$(\partial S/\partial V)_T = nR/V$$

Differentiating with respect to T at constant V we have

$$\frac{\partial^2 S}{\partial T \partial V} = \frac{\partial}{\partial T}(nR/V)_V = 0 \qquad ...(ix)$$

When V is constant dV = 0, and

$$\left(\frac{\partial S}{\partial T}\right)_V = (CV/T) \qquad ...(x)$$

On differentiation with respect to V at constant T, equation (x) gives

$$\frac{\partial^2 S}{\partial V \partial T} = \frac{\partial}{\partial V}(C_V/T)_T = 0 \qquad ...(xi)$$

From equation (ix) and (xi) we see that

$$\frac{\partial^2 S}{\partial T \partial V} = \frac{\partial^2 S}{\partial V \partial T}$$

Hence dS is an exact differentiai and S is a state function.

S *is a state function,* we have shown that

$$\int_1^2 \frac{dq_{rev}}{T} = C_V\ \text{In}\ (T_2/T_1) + nR\ \text{In}\ (V_2/V_1)$$

The right hand side of this equation depends only on the state variables T_1, V_1 and T_2, V_2, therefore the quantity on left side should also be a function only of state variables. This is

$$\int_1^2 \frac{dq_{rev}}{T} = \int_1^2 dS = S_2(T_2, V_2) - S_1(T_1, V_1) = \Delta S \qquad ...(xii)$$

$$\therefore\ \Delta S = C_V\ \text{In}\ (T_2/T_1) + nR\ \text{In}\ (V_2/V_1)$$

This state function S is called entropy function and like other thermodynamic functions the changes in entropy depend only on the state of the system and are independent of the path followed.

Equations (vi) and (xii) are the mathematical statements of the second law of thermodynamics and are used to define the entropy function. Here it must be regarded that this mathematical concept of entropy ($dS = dq_{rev}/T$) holds only for a *Reversible* change and change for an irreversible change $dS \neq dq_{irr}/T$.

GAS EQUATION

Consider 1 gram of the working substance (ideal gas) perfectly insulated from the surroundings. Let the external work done by the gas be δW.

Applying the first law of thermodynamics

$$\delta H = dU + \delta W$$

But $$\delta H = 0$$

and $$\delta W = P.dV$$

where P is the pressure of the gas and dV is the change in volume.

$$\therefore \quad 0 = dU + \frac{P.dV}{J}$$

As the external work Is done by the gas at the cost of its internal energy, there is fall in temperature by dT.

$$dU = 1 \times C_v \times dT$$

$$C_V.dT + \frac{P.dV}{J} = 0 \quad \text{...(ii)}$$

For an ideal gas*

$$PV = rT \quad \text{...(iii)}$$

Differentiating,

$$P.dV + V.dP = r.dT$$

Substituting the value of dT in equation (ii),

$$C_V\left[\frac{P.dV + V.dP}{r}\right] + \frac{P.dV}{J} = 0$$

$$C_V[P.dV + V.dP] + r.\frac{P.dV}{J} = 0$$

But, $$\frac{r}{J} = C_P - C_V$$

$\therefore \quad C_v.P.dV + C_v.V.dP + C_p.PdV - C_vPdV = 0$

$C_p.P.dV + C_v.V.dP = 0$

Integrating, $C_v.PV$,

$$\frac{C_P}{C_V}.\frac{dV}{V} + \frac{dP}{P} = 0$$

But $$\frac{C_P}{C_V} = 0$$

$$\therefore \quad \frac{dP}{P} + \gamma\frac{dV}{V} = 0$$

Integrating, $\log P + \gamma \log V = \text{const.}$

or $$\log PV^\gamma = \text{const.}$$

$$PV^\gamma = \text{const.} \qquad ...(iv)$$

This is the equation connecting pressure and volume during an adiabatic process.

Taking $$PV = rT$$

or $$P = \frac{rT}{V}$$

$$\left(\frac{rT}{V}\right).V^\gamma = \text{cosnt.}$$

But r is const.

$$rTV^{\gamma-1} = \text{cosnt.}$$

$$\therefore \quad rTV^{\gamma-1} = \text{const.} \qquad ...(v)$$

Also $$V = \frac{rT}{P}$$

$$P\left[\frac{rT}{P}\right]^\gamma = \text{cosnt.}$$

or $$\frac{r^\gamma T^\gamma}{P^{\gamma-1}} = \text{const.}$$

or $$\frac{P^{\gamma-1}}{T^\gamma} = \text{const.} \qquad ...(vi)$$

Thus, during an adiabatic process

(i) PV_{γ} = const.

(ii) $TV^{\gamma-1}$ = const. and

(iii) $\dfrac{P^{\gamma-1}}{T^{\gamma}}$ = const.

SOME SOLVED PROBLEMS

Problem 1:

1 g of water at 20°C is converted into ice at –10°C at constant pressure. Heat capacity for 1 g of water is 4.2 J/g-K and that of ice is 2.1 J/g-K. Beat of fusion of ice at 0°C = 335 J/g. Calculate the total change in the entropy of the system.

Solution:

(i) Change in entropy when the temperature of 1 g of water at 293 K falls to 273 K.

$$dS = \frac{\delta H}{T} = ms\int_{T_1}^{T_2}\frac{dT}{T}$$

$$= 1 \times 4.2\int_{293}^{273}\frac{dT}{T}$$

$$= 4.2 \times 2.3026 \log_{10}\left(\frac{273}{293}\right)$$

$$= \mathbf{-0.2969\ J/K.}$$

(ii) Change in entropy when 1 g of water at 273 K is converted into ice at 273 K

$$dS = \frac{\delta H}{T} = \frac{-1 \times 335}{273} = -1.227 \text{ J/K}$$

(iii) Change in entropy when the temperature of 1 g of ice at 273 K falls to 263 K

$$dS = \frac{\delta H}{T} = ms\int_{T_1}^{T_2}\frac{dT}{T}$$

$$= 1 \times 2.1 \times 2.3026 \log_{10}\left(\frac{263}{273}\right)$$

$$= -0.07834 \text{ J/K}$$

Total change in entropy of the system

$= -0.2969 - 1.227 - 0.07834$

$= \mathbf{-1.60224\ J/K.}$

Negative sign shows that there is decrease in entropy of the system.

Problem 2:

1 kg of wafer of 273 K is brought in contact with a heat reservoir at 873 K:

(1) what is the change in entropy of water when its temperature reaches 873 K?

(2) What is the change in entropy of (i) the reservoir and (ii) the universe.

Solution:

(1) Increase in entropy when the temperature of 1000 g of water is raised from 273 K to 373 K

$$dS = \int_{T_1}^{T_2} \frac{\delta H}{T}$$

$$= ms \times 2.3026 \log_{10} T_2/T_1$$

$$= 1000 \times 1 \times 2.3026 \log_{10} 373/273$$

$$= \mathbf{312\ cal/K.}$$

(2) (i) Change in entropy of the reservoir,

$$dS = \frac{-\delta H}{T}$$

$$= -\frac{1000 \times 1 \times 100}{372}$$

$$= \mathbf{-268.1\ cal/K.}$$

Negative sign shows decrease in entropy

(2) (ii) Change in entropy of the universe

$$= 312 - 268.1$$

$$= \mathbf{43.9\ cal/K.}$$

Therefore, the net increase in entropy of the universe

$$= \mathbf{43.9\ cal/K.}$$

Problem 3:

Deduce Clapeyron's latent heat equation from Maxwell's thermodynamical relations.

Solution:

We have $$\left(\frac{\partial H}{\partial V}\right)_T = T\left(\frac{\partial P}{\partial T}\right)_V \qquad ...(i)$$

Here $\left(\frac{\partial H}{\partial V}\right)_T$ represents the quantity of heat absorbed per unit increase in volume at constant temperature. This quantity of heat absorbed at constant temperature is the latent heat. If ∂H is the latent heat (L), when a unit mass of the substance changes in volume for V_1 to V_2 at constant temperature, then

$$\partial H = L \quad \text{and} \quad \partial V = V_2 - V_1$$

Substituting these values, in equation (i)

$$\therefore \qquad \left(\frac{L}{V_2 - V_1}\right)_T = T\left(\frac{\partial P}{\partial T}\right)_V$$

or $$\frac{L}{V_2 - V_1} = T\frac{dP}{dT}$$

or $$\frac{dP}{dT} = \frac{L}{T(V_2 - V_1)} \qquad ...(ii)$$

This is Clapeyron's latent heat equation.

Problem 4:

Show that for a homogeneous fluid

$$C_P - C_V = T\left(\frac{\partial P}{\partial T}\right)_V \left(\frac{\partial V}{\partial T}\right)_P$$

Also show that for a perfect gas $C_P - C_V = R$ and for a gas obeying Van der Waah equation

Solution:

We have $$C_P - C_V = R\left(1 + \frac{2a}{RTV}\right)$$

Here, $C_P \left(\frac{\partial H}{\partial T}\right)_P = T\left(\frac{\partial S}{\partial T}\right)_P$

and $C_V \left(\frac{\partial H}{\partial T}\right)_V = T\left(\frac{\partial S}{\partial T}\right)_V$

Considering S as a function of temperature and volume

$$dS = \left(\frac{\partial S}{\partial T}\right)_V = dT\left(\frac{\partial S}{\partial T}\right)_T dV$$

$$\left(\frac{\partial S}{\partial T}\right)_P = \left(\frac{\partial S}{\partial T}\right)_V \left(\frac{\partial T}{\partial T}\right) + \left(\frac{\partial S}{\partial V}\right)_T \left(\frac{\partial V}{\partial T}\right)_P$$

But from equation (v) [§ 6.52] $\left(\frac{\partial S}{\partial V}\right)_T = \left(\frac{\partial P}{\partial T}\right)_V$

$$T\left(\frac{\partial S}{\partial T}\right)_P = T\left(\frac{\partial S}{\partial T}\right)_V + T\left(\frac{\partial P}{\partial T}\right)_V \left(\frac{\partial V}{\partial T}\right)_P$$

$$C_P = C_V + T\left(\frac{\partial P}{\partial T}\right)_V \left(\frac{\partial V}{\partial T}\right)_P$$

$\therefore$ $$C_P - C_V = T\left(\frac{\partial P}{\partial T}\right)_V \left(\frac{\partial V}{\partial T}\right)_P \quad ...(i)$$

For a Perfect Gas

$$PV = RT$$

$\therefore$ $$\left(\frac{\partial P}{\partial T}\right)_V = \frac{R}{V} \text{ and } \left(\frac{\partial V}{\partial T}\right)_P = \frac{R}{P}$$

But, $C_P - C_V = T\left(\frac{\partial P}{\partial T}\right)_V \left(\frac{\partial V}{\partial T}\right)_P$

$$C_P - C_V = T\left(\frac{R}{V}\right)\left(\frac{R}{P}\right)$$

$$C_P - C_V = \frac{RT^2}{PV} = \frac{R^2T}{RT} = R$$

Hence $C_P - C_V = R$

[R is in heat units]

For a gas obeying Van der Waals equation

$$\left(P + \frac{a}{V^2}\right)(V - b) = RT$$

$$\left(P + \frac{a}{V^2}\right) = \frac{RT}{(V - b)}$$

$$\therefore \quad \left(\frac{\partial P}{\partial T}\right)_V = \frac{R}{V - b}$$

and $\left(\frac{-2a}{V^3} + \frac{RT}{(V - b)^2}\right)\left(\frac{\partial V}{\partial T}\right)_P = \frac{R}{V - b}$

But $C_P - C_V = T\left(\frac{\partial P}{\partial T}\right)_V \left(\frac{\partial V}{\partial T}\right)_P$

$$C_P - C_V = \frac{T\left(\frac{R}{V - b}\right)\left(\frac{R}{V - b}\right)}{\left[\frac{RT}{(V - b)^2} - \frac{2a}{V^3}\right]}$$

$$C_P - C_V = \frac{RT\,(V - b)^2}{RT\,(V - b)^2}\left[\frac{R}{1 - \frac{2a(V - b)}{V^3 RT}}\right]$$

Neglecting 6 as compared to V,

$$C_P - C_V = \frac{R}{\left(1 - \frac{2a}{VRT}\right)}$$

Since a is also small as compared to V

$$C_P - C_V = R\left(1 + \frac{2a}{VRT}\right) \qquad ...(iii)$$

[R is in heat units.]

Problem 5:

Derive the specific heat relation or show that

$$C_P - C_V = -T\left(\frac{\partial V}{\partial T}\right)_P^2 \left(\frac{\partial P}{\partial V}\right)_T$$

and $C_p - C_V = -TEa^2V$

where T is the absolute temperature, Sthe bulk modulus of elasticity, a the coefficient of volume expansion and V the specific volume,

Solution:

We have $$C_P\left(\frac{\partial H}{\partial T}\right) = T\left(\frac{\partial S}{\partial T}\right)_P$$

and $$C_V\left(\frac{\partial H}{\partial T}\right)_V = T\left(\frac{\partial S}{\partial T}\right)_V$$

Considering S as a function of temperature and volume,

$$dS\left(\frac{\partial S}{\partial T}\right)_V = dT + \left(\frac{\partial S}{\partial V}\right)_T dV$$

$$\left(\frac{\partial S}{\partial T}\right)_P = \left(\frac{\partial S}{\partial T}\right)_V\left(\frac{\partial T}{\partial T}\right) + \left(\frac{\partial S}{\partial V}\right)_T\left(\frac{\partial V}{\partial T}\right)_P$$

But from equation (v) of article 6.52,

$$\left(\frac{\partial S}{\partial V}\right)_T = \left(\frac{\partial P}{\partial T}\right)_V$$

$$\therefore \quad T\left(\frac{\partial S}{\partial V}\right)_P = T\left(\frac{\partial S}{\partial T}\right)_V + T\left(\frac{\partial P}{\partial T}\right)_V\left(\frac{\partial V}{\partial T}\right)_P$$

$$C_P = C_V + T\left(\frac{\partial P}{\partial T}\right)_V\left(\frac{\partial V}{\partial T}\right)_P \qquad \text{...(i)}$$

Taking the general equation of state for a gas as

$$P = f(V, T)$$

$$C_P = C_V + T\left(\frac{\partial P}{\partial T}\right)_V\left(\frac{\partial V}{\partial T}\right)_P$$

$$\frac{dP}{dT} = \left(\frac{\partial P}{\partial T}\right)_V + \left(\frac{\partial P}{\partial V}\right)_T\left(\frac{\partial V}{\partial T}\right)_P$$

At constant pressure,

$$dP = 0$$

$$\therefore \quad \left(\frac{\partial P}{\partial T}\right)_V = -\left(\frac{\partial P}{\partial V}\right)_T\left(\frac{\partial V}{\partial T}\right)_P$$

Substituting this value in eaquation (i)

$$C_P - C_V = -\left(\frac{\partial P}{\partial V}\right)_T \left(\frac{\partial V}{\partial T}\right)_P^2 \qquad ...(ii)$$

$$C_P - C_V = -T\left(\frac{\partial V}{\partial T}\right)_P^2 \left(\frac{\partial P}{\partial V}\right)_T \qquad ...(iii)$$

But $E = V\left(\frac{\partial P}{\partial V}\right)_T$

and $\alpha = \frac{1}{V}\left[\frac{\partial V}{\partial T}\right]_P$

Substituting thee values in equation (ii)

$$C_P - C_V = -TEa^2V. \qquad ...(iv)$$

Problem 6:

Show that $C_2 - C_1 = \frac{dL}{dT} - \frac{L}{T}$

where C_1 and C_2 represent the specific heta of a liquid and it saturted vapour and l is the latent heat of the vapour.

Solution:

For a change of state from liquid to vapour,

$$S_2 - S_1 = \frac{L}{T} \qquad ...(i)$$

Here S_1 and S_2 are the entropies in the liquid and vapour states respectively.

Differentiating equation (i) with respect to T

$$\frac{dS_2}{dT} - \frac{dS_1}{dT} = -L.\frac{1}{T^2} + \frac{1}{T}.\frac{dL}{dT}$$

$$T\left(\frac{dS_2}{dT}\right) - T\left(\frac{dS_1}{dT}\right) = -\frac{L}{T} + \frac{dL}{dT}$$

$$C_2 - C_1 = \frac{dL}{dT} - \frac{L}{T}$$

This equation is known an Clausius latent heat equation.

Problem 7:

Calculate under what pressure water would boil at 160°C if the change in specific volume when 1 gram of water is converted into steam is 1676 cc. Given latent heat of vaporisation of steam = 540 cal per gram; J = 4.2 × 10^7 ergs/cal and one atmosphere pressure = 10^6 dynes/cm^2.

Solution:

Here $\delta H = 540 \text{ cal} = 540 \times 4.2 \times 107 \text{ erga}$

$\delta V = 1676 \text{ cc}$

$T = 373 \text{ K, and } 150°C = 423 \text{ K}$

$\delta T = 423 - 373 = 50 \text{ K}$

$\delta P = ?$

Applying these values in the Maxwell's thermodynamical relation

$$\left(\frac{\delta H}{\partial V}\right)_T = T\left(\frac{\partial P}{\partial T}\right)_V$$

$$\frac{540 \times 4.2 \times 10^7}{1676} = 373\left(\frac{\partial P}{50}\right)$$

$$\partial P = \frac{540 \times 4.2 \times 10^7 \times 50}{373 \times 1676}$$

$$= 1.814 \times 10^6 \text{ dynes/cm}^2$$

$$= 1.814 \text{ atmospheres}$$

Therefore, the pressure at which water would boil at 150°C

$$= 1.814 + 1.000$$

$$= \mathbf{2.814 \text{ atmospheres.}}$$

Problem 8:

Calculate under what pressure ice would freeze at –1°C, if the change in specific volume when 1 gram of water freezes into ice is 0.091 cc.

Given latent heat of fusion of ice = 80 cal/g, J = 4.2 × 10^7 ergs/cal and one atmosphere pressure = 10^6 dynes/cm^3.

Solution:

Here, $\delta H = 80 \text{ cal} + 80 \times 4.2 \times 10^7 \text{ ergs}$

$\partial V = 0.091 \text{ cc}$

$T = 273 \text{ K}$

$\partial T = 273 - 272 = 1\text{K}$

$\partial P = ?$

Applying these values in the Maxwell's thermodynamical relation,

$$\left(\frac{\delta H}{\partial V}\right)_T = T\left(\frac{\partial P}{\partial T}\right)_V$$

$$\frac{80 \times 4.2 \times 10^7}{0.091} = \frac{273 \times \partial P}{1}$$

$$\partial P = 135.2 \times 10^6 \text{ dyne/cm}^2$$

$$= 135.2 \text{ atmospheres}$$

Therefore, the pressure under which ice would freeze = atmospheric pressure + $\partial P = 1 + 135.2 =$ **136.2 atmospheres.**

Problem 9:

Calculate the specific heat of saturated steam given that the specific heat of water at 100°C = 1.01 and latent heat of vaporisation decreases with rise in temperature at the rate of 0.64 cal/K. Latent heat of vaporization of steam = 540 cal.

Solution:

We have $C_1 = 1.01$

$C_2 = ?$

$T = 373 \text{ K}$

$$\frac{dL}{dT} = -0.64 \text{ cal/K}$$

$$C_2 - C_1 = \frac{dL}{dT} - \frac{L}{T}$$

$$C_2 = C_1 + \frac{dL}{dT} - \frac{L}{T}$$

$$= 1.01 + (-0.64) - 540/373$$

$$= -1.077 \text{ cal/g.}$$

It shows that the specific heat of saturated steam is negative.

Problem 10:

Calculate the specific heat of saturated steam at 100°C from the following data:

L at 90°C + 545.25 cal

L at 100°C = 539.30 cal

L at 110°C = 533.17 cal

Specific heat of water at 100°C = 1.013 cal/g.

Solution:

Here $$\frac{dL}{dT} = \frac{533.17 - 545.25}{20}$$

$$= \frac{-12.08}{20} = -0604 \text{ cal/K}$$

$$L = 539.3 \text{ cal}$$

$$T = 373 \text{ K}$$

$$C = 1.013$$

$$C = ?$$

$$C_2 - C_1 = \frac{dL}{dT} - \frac{L}{T}$$

$$C_2 = C_1 + \frac{dL}{dT} - \frac{L}{T}$$

$$C_2 = 1.013 - 0.604 - \frac{539.3}{373}$$

$$C_2 = -1.036 \text{ cal/g.}$$

Problem 11:

Using Maxwell's thermodynamical relations, prove that for any substance, the ratio of the adiabatic and isothermal elasticities it equal to the ratio of the two specific heats.

Solution:

From definition,

Isothermal elasticity $E_T = -V\left(\frac{\partial P}{\partial V}\right)_T$

Adiabatic elasticity $E_T = -V\left(\frac{\partial P}{\partial V}\right)_S$

$$\therefore \qquad \frac{E_S}{E_T} = \frac{-V\left(\frac{\partial P}{\partial V}\right)_S}{-V\left(\frac{\partial P}{\partial V}\right)_T}$$

$$= \frac{\left(\frac{\partial P}{\partial V}\right)_S}{\left(\frac{\partial P}{\partial V}\right)_T}$$

$$= \frac{\left(\frac{\partial P}{\partial T}\right)_S\left(\frac{\partial T}{\partial V}\right)_S}{\left(\frac{\partial P}{\partial S}\right)_T\left(\frac{\partial S}{\partial V}\right)_T}$$

$$\frac{E_S}{E_T} = \frac{\left(\frac{\partial S}{\partial P}\right)_T\left(\frac{\partial T}{\partial V}\right)_S}{\left(\frac{\partial T}{\partial P}\right)_S\left(\frac{\partial S}{\partial V}\right)_T} \qquad \text{...(i)}$$

But from the first four Maxwell s equations

$$\left(\frac{\partial S}{\partial P}\right)_T = -\left(\frac{\partial V}{\partial T}\right)_P$$

$$\left(\frac{\partial T}{\partial V}\right)_S = -\left(\frac{\partial P}{\partial S}\right)_V$$

$$\left(\frac{\partial T}{\partial P}\right)_S = \left(\frac{\partial V}{\partial S}\right)_P$$

$$\left(\frac{\partial S}{\partial V}\right)_T = \left(\frac{\partial P}{\partial T}\right)_V$$

$\therefore$ Substituting these values in equation (i)

$$\frac{E_S}{E_T} = \frac{\left(\frac{\partial V}{\partial T}\right)_P\left(\frac{\partial P}{\partial S}\right)_V}{\left(\frac{\partial V}{\partial S}\right)_P\left(\frac{\partial P}{\partial T}\right)_V}$$

$$\frac{E_S}{E_T} = \frac{\left(\frac{\partial S}{\partial V}\right)_P \left(\frac{\partial V}{\partial T}\right)_P}{\left(\frac{\partial S}{\partial P}\right)_V \left(\frac{\partial P}{\partial T}\right)_V}$$

$$= \frac{\left(\frac{\partial S}{\partial T}\right)_P}{\left(\frac{\partial S}{\partial T}\right)_V}$$

$$= \frac{\left(\frac{\partial H}{\partial T}\right)_P}{\left(\frac{\partial H}{\partial T}\right)_V}$$

But $\left(\frac{\partial H}{\partial T}\right)_P = C_P$

and $\left(\frac{\partial H}{\partial T}\right)_V = C_V$

$$\therefore \quad \frac{E_S}{E_T} = \frac{C_P}{C_V} = \gamma.$$

Problem 12:

Using MavuxIFs thermodynamical relatwns, prove that the ratio of the adiabatic to the isobaric coefficient of expansion is

$$\frac{1}{(1 - \gamma)}$$

Solution:

Adiabatic volume coefficient of expansion,

$$\alpha_2 = \frac{1}{V}\left(\frac{\partial V}{\partial T}\right)_S \qquad ...(i)$$

Isobaric volume coefficient of expansion,

$$\alpha_P = \frac{1}{V}\left(\frac{\partial V}{\partial T}\right)_F \qquad ...(ii)$$

$$\therefore \quad \frac{\alpha_S}{\alpha_P} = \frac{\frac{1}{V}\left(\frac{\partial V}{\partial T}\right)_S}{\frac{1}{V}\left(\frac{\partial V}{\partial T}\right)_P} \qquad ...(iii)$$

$$= \frac{\left(\frac{\partial V}{\partial T}\right)_S}{\left(\frac{\partial V}{\partial T}\right)_P} = \frac{1}{\left(\frac{\partial T}{\partial V}\right)_S \left(\frac{\partial V}{\partial T}\right)_P}$$

But $\left(\frac{\partial T}{\partial V}\right)_S = -\left(\frac{\partial P}{\partial S}\right)_V$

$$\therefore \qquad \frac{\alpha_S}{\alpha_P} = -\frac{1}{\left(\frac{\partial P}{\partial S}\right)_V \left(\frac{\partial V}{\partial T}\right)_P}$$

$$= \frac{1}{-\left[\left(\frac{\partial P}{\partial T}\right)_V \left(\frac{\partial T}{\partial S}\right)_V \left(\frac{\partial V}{\partial T}\right)_P\right]}$$

$$= \frac{\left(\frac{\partial S}{\partial T}\right)_V}{-\left(\frac{\partial P}{\partial T}\right)_V \left(\frac{\partial V}{\partial T}\right)_P}$$

$$= \frac{T\left(\frac{\partial S}{\partial T}\right)_V}{-T\left(\frac{\partial P}{\partial T}\right)_V \left(\frac{\partial V}{\partial T}\right)_P}$$

But $C_P - C_V = T\left(\frac{\partial P}{\partial T}\right)_V \left(\frac{\partial V}{\partial T}\right)_P$

and $\qquad C_V = T\left(\frac{\partial S}{\partial T}\right)_V$

$$\therefore \qquad \frac{\alpha_S}{\alpha_P} = \frac{C_V}{-(C_P - C_V)}$$

or

$$\frac{\alpha_S}{\alpha_P} = \frac{1}{1 - \left(\frac{C_P}{C_V}\right)}$$

or

$$\frac{\alpha_S}{\alpha_P} = \frac{1}{1-\gamma} \qquad \text{...(iv)}$$

Problem 13:

Using Maxwell's themwdynamical relations, prove that the ratio of the adiabatic to the isochoric pressure coefficient of expansion is equal to

$$\frac{\gamma}{(\gamma - 1)}$$

Solution:

Adiabatic prenxire coefficient of expansion,

$$\beta_S = \frac{1}{P}\left(\frac{\partial P}{\partial T}\right)_S \qquad \text{...(i)}$$

Isochoric pressure coefficient of expanison.

$$\beta_V = \frac{1}{P}\left(\frac{\partial P}{\partial T}\right)_V \qquad \text{...(ii)}$$

$$\therefore \qquad \frac{\beta_S}{\beta_V} = \frac{\frac{1}{P}\left(\frac{\partial P}{\partial T}\right)_S}{\frac{1}{P}\left(\frac{\partial P}{\partial T}\right)_V} \qquad \text{...(iii)}$$

$$= \frac{\left(\frac{\partial P}{\partial T}\right)_S}{\left(\frac{\partial P}{\partial T}\right)_V}$$

$$= \frac{1}{\left(\frac{\partial T}{\partial P}\right)_S \left(\frac{\partial P}{\partial T}\right)_V}$$

But $\left(\frac{\partial T}{\partial P}\right)_S = \left(\frac{\partial V}{\partial S}\right)_P$ ...(iv)

$$\frac{\beta_S}{\beta_V} = \frac{1}{\left(\frac{\partial V}{\partial S}\right)_P \left(\frac{\partial P}{\partial T}\right)_V} \qquad \text{...(v)}$$

$$= \frac{1}{\left(\frac{\partial V}{\partial T}\right)_P \left(\frac{\partial T}{\partial S}\right)_P \left(\frac{\partial P}{\partial T}\right)_V}$$

$$= \frac{\left(\frac{\partial S}{\partial T}\right)_P}{\left(\frac{\partial V}{\partial T}\right)_P \left(\frac{\partial P}{\partial T}\right)_V}$$

$$= \frac{T\left(\frac{\partial S}{\partial T}\right)_P}{T\left(\frac{\partial V}{\partial T}\right)_P \left(\frac{\partial P}{\partial T}\right)_V}$$

$$= \frac{C_P}{C_P - C_V}$$

or $$\frac{\beta_S}{\beta_V} = \frac{\left(\frac{C_P}{C_V}\right)}{\left(\frac{C_P}{C_V} - 1\right)} = \frac{\gamma}{\gamma - 1} \qquad ...(vi)$$

Problem 14:

Using Maxwell's thermodynamical relations, show that

$$\left(\frac{\partial C_V}{\partial V}\right) = T\left(\frac{\partial^2 S}{\partial V.\partial T}\right) = T\left(\frac{\partial^2 P}{\partial T^2}\right)_V$$

Solution:

From Maxwell's relations,

$$\left(\frac{\partial S}{\partial V}\right)_T = \left(\frac{\partial P}{\partial T}\right)_V \qquad ...(i)$$

Differentiating equation (i) with respect to temperature,

$$\left(\frac{\partial^2 S}{\partial T.\partial V}\right) = \left(\frac{\partial^2 P}{\partial T^2}\right)_V \qquad ...(ii)$$

But $$C_V = T\left(\frac{\partial S}{\partial T}\right)_V \qquad ...(iii)$$

Differentiating equation (iii) with respect to volume,

$$\left(\frac{\partial C_V}{\partial V}\right) = T\left(\frac{\partial^2 S}{\partial T.\partial V}\right) \qquad ...(iv)$$

From equations (ii) and (iv)

$$\left(\frac{\partial C_V}{\partial V}\right) = T\left(\frac{\partial^2 S}{\partial T.\partial V}\right) = T\left(\frac{\partial^2 P}{\partial T^2}\right)_V \qquad ...(v)$$

Problem 15:

Using Maxwell's thermodynamical relations, show that

$$\left(\frac{\partial C_P}{\partial P}\right) = T\left(\frac{\partial^2 S}{\partial P.\partial T}\right) = -T\left(\frac{\partial^2 V}{\partial T^2}\right)_P$$

Solution:

From Maxwell's relations,

$$\left(\frac{\partial S}{\partial P}\right)_T = -\left(\frac{\partial V}{\partial T}\right)_P \qquad ...(i)$$

Differentiating equation (i) with respect to temperature,

$$\left(\frac{\partial^2 S}{\partial T.\partial P}\right) = -\left(\frac{\partial^2 V}{\partial T^2}\right)_P \qquad ...(ii)$$

But $$C_P = T\left(\frac{\partial S}{\partial T}\right)_P \qquad ...(iii)$$

Differentiating equation (iii) with respect to pressure,

$$\left(\frac{\partial C_P}{\partial P}\right) = T\left(\frac{\partial^2 S}{\partial T.\partial P}\right) \qquad ...(iv)$$

From equations (ii) and (iv)

$$\left(\frac{\partial C_P}{\partial P}\right) = T\left(\frac{\partial^2 S}{\partial T.\partial P}\right) = -T\left(\frac{\partial^2 V}{\partial T^2}\right)_P \qquad ...(v)$$

Problem 16:

Show that for a perfect gas

$$\left(\frac{\partial U}{\partial V}\right)_T = 0$$

Solution:

From the Maxwell's first equation,

$$\left(\frac{\partial S}{\partial V}\right)_T = \left(\frac{\partial P}{\partial T}\right)_V$$

$$\frac{1}{T}\left(\frac{\partial H}{\partial V}\right)_T = \left(\frac{\partial P}{\partial T}\right)_V$$

or $$\left(\frac{\partial H}{\partial V}\right)_T = T\left(\frac{\partial P}{\partial T}\right)_V$$

But $\delta H = dU + PdV$

$$\therefore \quad \left(\frac{dU + PdV}{dV}\right) = T\left(\frac{\partial P}{\partial T}\right)_V$$

$$\frac{dU}{dV} + P = T\left(\frac{\partial P}{\partial T}\right)_V$$

$$\left(\frac{dU}{dV}\right)_T = T\left(\frac{\partial P}{\partial T}\right)_V - P$$

But from perfect gas equation,

$$PV = RT$$

$$\therefore \quad \left(\frac{\partial P}{\partial T}\right)_V = \frac{R}{V}$$

$$\left(\frac{\partial U}{\partial V}\right)_T = T\left(\frac{R}{V}\right) - P$$

or $$\left(\frac{\partial U}{\partial V}\right)_T = P - P = 0$$

$$\therefore \quad \left(\frac{\partial U}{\partial V}\right)_T = 0.$$

Problem 17:

When a system is taken from the state A to the state B, along the path ACB, 80 joules of heat flows into the system, and the system does 30 joules of work.

(a) Bow much heat flows into the system along the path ADB, if the work done is 10 joules.

(b) The system is returned from the state B to the state A along the curved path. The work done on the system is 20 joules. Does the system absorb or liberate heat and how much?

(c) If $U_A = 0$, $U_D = 40$ joules, find the heat absorbed in the process AD and DB.

Solution:

Along the path ACB,

$$HACB = U_B - U_A + W$$

Here $H = +80$ joules

$W = +30$ joules

$$\therefore +80 = U_B - U_A + 30$$

$$U_B - U_A = 80 - 30 = 50 \text{ joules}$$

(a) Along the path ADB,

$$W = +10 \text{ joules}$$

$$H_{ADB} = U_B - U_A + W$$

$$H = 50 + 10 = 60 \text{ joules}$$

(b) For the curved path from B to A,

$$W = -20 \text{ joules}$$

$$H = (U_A - U_B) + W$$

$$= -50 - 20 = -70 \text{ joules}$$

(–ve sign shows that heat is liberated by the system)

(c) $U_A = 0$,

$$U_D = 40 \text{ joules}$$

$$U_B - U_A = 50$$

$$\therefore \quad U_B = 50 \text{ joules}$$

In the process ADB, 10 joules of work is done. Work done from 4 to D is +10 joules and from D to B is zero.

For AD,

$$H_{AD} = (U_D - U_A) + W$$

$$= 40 + 10 = 50 \text{ joules}$$

For DB

$$H_{DB} = U_C - U_D + W$$

$$= \mathbf{50 - 40 + 0 = 10 \text{ joules.}}$$

Problem 18:

A motor car tyre has a pressure of 2 atmospheres at the room temperature of 27°C. If the tyre suddenly bursts, find the resulting temperature.

Solution:

Here,

$$P_1 = 2 \text{ atmospheres}$$

$$T_1 = 273 + 27$$

$$= 300 \text{ K}$$

$$P_2 = 1 \text{ atmosphere}$$

$$T_2 = ?$$

$$\gamma = 1.4$$

$$\frac{P_1^{\gamma-1}}{T_1^{\gamma}} = \frac{P_2^{\gamma-1}}{T_2^{\gamma}}$$

$$\left(\frac{P_2}{P_1}\right)^{\gamma-1} = \left(\frac{T_2}{T_1}\right)^{\gamma}$$

$$\left(\frac{1}{2}\right)^{0.4} = \left(\frac{T_2}{300}\right)^{1.4}$$

$$0.4 \log (0.5) = 1.4 [\log T_2 - \log 300]$$

$$-0.1204 = 1.4 \log T_2 - 3.4680$$

$$1.4 \log T_2 = 3.4680 - 0.1204$$

$$= 3.3476$$

$$\log T_2 = \frac{3.3476}{1.4}$$

$$= 2.3911$$

$$T_2 = 246.1 \text{ K}$$

$$= \mathbf{-26.9°C.}$$

Problem 19:

A quantity of air at 27°C and atmospheric Pressure it suddenly compressed to half its original volume, find the final (i) pressure and (ii) temperature.

Solution:

(i) $P_1 - 1$ atmosphere; $P_2 = ?$, $\gamma = 1.4$

$$V_1 = V;$$

$$V_2 = V/2$$

During sudden compression, the process is adiabatic

$$P_1V_1^{\gamma} = P_2V_2^{\gamma}$$

$$P_2 = P_1\left[\frac{V_1}{V_2}\right]^{\gamma}$$

$$= 1[2]^{1/4}$$

$$= 2.636 \text{ atmospheres}$$

(ii) $V_1 = V;$

$$V_2 = V/2$$

$$T_1 = 300\text{ K};$$

$$T_2 = ?$$

$$\gamma = 1.4$$

$$T_1(V_1)^{\gamma-1} = T_2\,(V_2)^{\gamma-1}$$

$$T_2 = T_1[2]^{1/4-1}$$

$$= 300[2]^{0.4}$$

$$= 395.9\text{ K}$$

$$= \mathbf{122.9°C.}$$

Problem 20:

Air is compressed adiabatically to half its volume. Calculate the change in its temperature.

Solution:

Let the initial temperature be T_1 K and the final temperature

Initial volume $= V_1$

Final volume$= V_2$

$= V_1/2$

During an adiabatic process

$$T_1 V_1^{\gamma-1} = T_2 V_2^{\gamma-1}$$

$$T_2 = T_1 \left[\frac{V_1}{V_2}\right]^{\gamma-1}$$

$$T_2 = T_1[2]^{\gamma-1}$$

But γ for air $= 1.40$

$$T_2 = T_1[2]^{1.40-1}$$

$$T_2 = T_1[2]^{0.40}$$

$$T_2 = 1.319\ T_1$$

Change in temperature

$$= T_2 - T_1$$

$$= 1.319\ T_1 - T_1$$

$$= \mathbf{1.319\ T_1\ K.}$$

Problem 21:

1 gram molecule of a monoatomic (γ = 5/3) perfect gas at 27°C is adiabatically compressed in a reversible process from an initial pressure of 1 atmosphere to a final pressure of 60 atmospheres. Calculate the resulting difference in temperature.

Solution:

In a reversible adiabatic process

$$\frac{P_1^{\gamma-1}}{T_1^{\gamma}} = \frac{P_2^{\gamma-1}}{T_2^{\gamma}}$$

or $$\left(\frac{P_2}{P_1}\right)^{\gamma-1} = \left(\frac{T_2}{T_1}\right)^{\gamma}$$

Here, $P_2 = 50,$

$P_1 = 1,$

$$T_1 = 273 + 27$$
$$= 300 \text{ K}$$
$$T_2 = ?$$
$$\gamma = 5/3$$
$$\therefore \quad (50)^{2/3} = \left(\frac{T_2}{300}\right)^{5/3}$$
$$2/3 \log (50) = 5/3 [\log T_2 - \log 300]$$
$$\mathbf{T_2 = 1{,}434 \text{ K}}$$
$$\mathbf{= 1{,}161°C.}$$

Problem 22:

A quantity, of dry air at 27°C is compressed (i) slowly and (ii) suddenly to 1/3 of its volume, find the change in temperature in each case, assuming f to be 1.4 for dry air.

Solution:

(1) When the process is slow, the temperature of the system remains constant. Therefore, there is no change in temperature.

(2) When the compression is sudden, the process is adiabatic.

Here $V_1 = V,$

$$V_2 = V/3$$
$$T_1 = 300 \text{ K},$$
$$T_2 = ?$$
$$\gamma = 1.4$$
$$T_2(V_2)^{\gamma-1} = T_1[V_1]^{\gamma-1}$$
$$T_2 = T_1\left[\frac{V_1}{V_2}\right]^{\gamma-1}$$

or
$$T_2 = 300\left[\frac{3V}{V}\right]^{\gamma-1}$$
$$= 300 [3]1.4.1$$

or
$$T_2 = 465.5 \text{ K}$$

$= 192.5°C$

The temperature of air increases by

192.5–27 = 165.5°C or 165.5 K.

Problem 23:

A certain mass of gas at NTP is expanded to three times its volume under adiabatic conditions. Calculate the resulting temperature and pressure, γ for the gas is 1.40.

Solution:

(1) Here, $V_1 = V,$

$V_2 = 3V$

$T_1 = 273 \text{ K}$

$T_2 = ?$

$$T_1 V_1^{\gamma-1} = T_2 V_2^{\gamma-1}$$

or $$T_2 = T_1 \left[\frac{V_1}{V_2}\right]^{\gamma-1}$$

$$T_2 = 273 \left[\frac{1}{3}\right]^{1.4-1}$$

T = 276 K = –97°C.

(2) Here, $V_1 = V,$

$V_2 = 3V$

$P_1 = 1$ atmosphere,

$P_2 = ?$

$$P_1 V_1^{\gamma} = P_2 V_2^{\gamma}$$

$$P_2 = P_1 \left[\frac{V_1}{V_2}\right]^{\gamma}$$

or $$P_2 = 1\left(\frac{1}{3}\right)^{1.4}$$

P_2 = 0.2148 atmosphere.

Problem 24:

Find the efficiency of the carnot's engine working between the steam point and the ice point.

Solution:

We have $T_1 = 273 + 100 = 373$ K

$T_2 = 273 + 0 = 273$ K

$$\eta = 1 - \frac{T_2}{T_1}$$

$$= 1 - \frac{273}{373} = \frac{100}{373}$$

$$\text{\% efficiency} = \frac{100}{373} \times 100$$

$$= \mathbf{26.81\%.}$$

Problem 25:

Find the efficiency of a carnot's engine working between 127°C and 27°C.

Solution:

We have $T_1 = 273 + 127 = 400$ K

$T_2 = 273 + 27 = 300$ K

$$\eta = 1 - \frac{T_2}{T_1}$$

$$= 1 - \frac{300}{400} = 0.25$$

% efficiency $= \mathbf{25\%.}$

Problem 26:

A carnot's engine whose temperature of the source is 400 K takes 200 calories of heat at this temperature and rejects 150 calories of heat to the sink. What is the temperature of the sink? Also calculate the efficiency of the engine.

Solution:

We have $H_1 = 200$ cal;

$H_2 = 150$ cal

$T_1 = 400$ cal;

$T_2 = ?$

$$\frac{H_1}{T_1} = \frac{H_2}{T_2}$$

$$T_2 = T_2 \frac{H_2}{H_1} \times 400 = 300 \text{ K}$$

$$\eta = 1 - \frac{T_2}{T_1}$$

$$= 1 - \frac{300}{400} = 0.25$$

% efficiency = **25%**

Problem 27:

A Carnot's engine is operated between two reservoirs at temperatures of 450 K and 350 K. If the engine receives 1000 calories of heat from the source in each cycle, calculate the amount of heat rejected to the sink in each cycle. Calculate the efficiency of the engine and the work done by the engine in each cycle. (1 calorie = 4.2 joules).

Solution:

We have $T_1 = 450$ K;

$T_2 = 350$ K

$H_1 = 1000$ cal;

$H_2 = ?$

$$\frac{H_2}{H_1} = \frac{T_2}{T_1}$$

$$H_2 = H_1 \times \frac{T_2}{T_1}$$

$$= \frac{1000 \times 350}{450} = 777.77 \text{ cals}$$

$$\eta = 1 - \frac{T_2}{T_1}$$

$$= 1 - \frac{350}{450} = \frac{100}{450}$$

$= 0.2222$

% efficiency $=$ **22.22%.**

Work done in each cycle

$= H_1 - H_2$

$= 1000 - 777.77$

$= 222.23$ cal

$= 222.23 \times 4.2$ joules

$=$ **933.33 Joules.**

Problem 28:

A carnot's engine working as a refrigerator between 260 k and 300 k receives 500 calories of heat from the reservoir at the lower temperature. Calculate the amount of heat rejected to the reservoir at the higher temperature. Calculate also the amount of work done in each cycle to operate the refrigerator.

Solution:

We have $H_1 = ?$

$H_2 = 500$ cal

$T_1 = 300$ K

$T_2 = 260$ K

$$\frac{H_1}{H_2} = \frac{T_1}{T_2};$$

$$H_1 = H_2 \cdot \frac{T_1}{T_2}$$

$$H_1 = \frac{500 \times 300}{260} = 576.92 \text{ cal}$$

$W = H_1 - H_2 = 76.92$ cal

$= 76.92 \times 4.2$ joules

$=$ **323.08 joules.**

Problem 29:

A carnot's refrigerator takes heat from water at 0°C and discards it to a room at 27°C. 1 kg of water at 0°C is to be changed into ice at

0°C. How many calories of heat are discarded to the room? What is the work done by the refrigerator in this process? What is the coefficient of performance of the machine?

Solution:

We have $H_1 = ?$

$$H_2 = 1000 \times 80 = 80{,}000 \text{ cal}$$

$$T_1 = 300 \text{ K}$$

$$T_2 = 273 \text{ K}$$

(1) $$\frac{H_1}{H_2} = \frac{T_1}{T_2}$$

$$H_1 = \frac{H_2 T_1}{T_2}$$

$$= \frac{80{,}000 \times 300}{273}$$

$$\mathbf{H_1 = 87{,}900 \text{ cal.}}$$

(2) Work done by the refrigerator

$$= W = J\,(H_1 - H_2)$$

$$W = 4.2\,(87{,}900 - 80{,}000)$$

$$W = 4.2 \times 7900$$

or $$W = 3.183 \times 10^4 \text{ joules}$$

(3) Coefficient of performance,

$$= \frac{H_2}{H_1 - H_2}$$

$$= \frac{80{,}000}{87{,}900 - 80{,}000}$$

$$= \frac{80{,}000}{7900}$$

$$= \mathbf{10.13.}$$

Problem 30:

A carnot engine whose low temperature reservoir is at 7°C has an efficiency of 50%. It is desired to increase the efficiency to 70%. By how many degrees should the temperature of the high temperature reservoir be increased?

Solution:

In the first case

$$\eta = 50\% = 0.5,$$

$$T_2 = 273 + 7 = 280 \text{ K}$$

$$T_1 = ?$$

$$\eta = 1 - \frac{T_2}{T_1}$$

or $$0.5 = 1 - \frac{280}{T_1}$$

or $$T_1 = 560 \text{ K}$$

In the second case

$$\eta' = 70\% = 0.7$$

$$T_2 = 280 \text{ K},$$

$$T_1' = ?$$

$$\eta' = 1 - \frac{T_2}{T_1'}$$

$$0.7 = 1 - \frac{280}{T_1'}$$

or $$T_1' = 840 \text{ K}$$

Increase in temperature = 840 – 560 = **280 K.**

Problem 31:

An inventor claims to have developed an engine working between 600 K and 300 K capable of having an efficiency of 52%, Comment on his claim.

Solution:

The efficiency of a Carnot's engine working between 600 K and 300 K,

$$\eta = 1 - \frac{T_2}{T_1}$$

$$= 1 - \frac{300}{600} = 0.5$$

$$= 50\%$$

The efficiency claimed = η = 52%.

It means that the efficiency of the engine is more than the efficiency of a Carnot's engine working between the same two temperature limits. But, no engine can have an efficiency more than a Carnot's engine, so his claim is invalid.

Problem 32:

In a double acting steam engine, the average pressure of steam is 10^5 newtons/metre^2. The length of the stroke if 1 metre and the area of the piston is 0.15 sq metre. Find the power of the engine, if it makes 5 strokes per second.

Solution:

Here P = Wnewtons/metre^2

L = 1 metre

A = 0.15 sq metre

N = 5 strokes/second

$$\text{Power} = \frac{2\,\text{PLAN}}{1000}$$

$$\text{Power of the engine} = \frac{2 \times 10^5 \times 1 \times 0.15 \times 5}{1000}$$

= **150 kilowatts.**

Problem 33:

Calculate the depression in the melting point of ice produced by one atmosphere increase of pressure. Given latent heat of ice = 80 cal per gram and the specific volumes of 1 gram of ice and water at 0°C are 1.091 cm^3 and 1.000 cm^3 respectively.

Solution:

Here L = 80 cal = 80 × 4.2 × 10^7 ergs

T = 273 K

$$dP = 1 \text{ atmosphere}$$
$$= 76 \times 13.6 \times 980 \text{ dynes/cm}^2$$
$$V_1 = 1.091 \text{ cm}^3$$
$$V_2 = 1.000 \text{ cm}^3$$
$$\frac{dP}{dT} = \frac{L}{T\,(V_2 - V_1)}$$
$$dT = \frac{dP.T.(V_2 - V_1)}{L}$$
$$= \frac{76 \times 13.6 \times 980 \times 273\,(1 - 1.091)}{80 \times 4.2 \times 10^7}$$
$$= 0.0074 \text{ K.}$$

Therefore, the decrease in the melting point of ice with an increase in pressure of one atmosphere

$$= \mathbf{0.0074\ K = 0.0074°C.}$$

Problem 34:

Find the increase in the boiling point of water at 100°C when the pressure is increased by one atmosphere. Latent heat of vaporisation of steam is 640 cal/gram and 1 gram of steam occupies a volume of 1677 cm³.

Solution:

We have $dP = 76 \times 13.6 \times 980 \text{ dynes/cm}^2$

$$T = 100 + 273$$
$$= 373 \text{ K}$$
$$L = 540 \times 4.2 \times 10^7 \text{ ergs}$$
$$V_1 = 1.000 \text{ cm}^3$$
$$V_2 = 1677 \text{ cm}^3$$
$$\frac{dP}{dT} = \frac{L}{T(V_2 - V_1)}$$
$$dT = \frac{dP \times T(V_2 - V_1)}{L}$$

$$= \frac{76 \times 13.6 \times 980 \times 373 \times 1676}{540 \times 4.2 \times 10^7}$$

$$= \mathbf{27.92°C.}$$

Therefore, the increase in the boiling point of water with an increase in pressure of one atmosphere

$$= \mathbf{27.92°C}$$

$$= \mathbf{27.92\ K.}$$

Problem 35:

Calculate the change in temperature of boiling water when the pressure is increased by 27.12 mm of Hg. The normal boiling point of water at atmospheric pressure is 100°C.

Latent heat of steam = *537 cal/g*

and specific volume of steam = *1674 cm³.*

Solution:

We have $dP = 2.712 \times 13.6 \times 980$ dynes/cm²

$$T = 100 + 273 = 373\ K$$

$$L = 537 \times 4.2 \times 10^7 \text{ ergs}$$

$$V_1 = 1.000 \text{ cm}^3$$

$$V_2 = 1674 \text{ cm}^3$$

$$\frac{dP}{dT} = \frac{L}{T(V_2 - V_1)}$$

$$dT = \frac{dP \times T\,(V_2 - V_1)}{L}$$

$$= \frac{2.712 \times 13.6 \times 980 \times 373 \times 1673}{537 \times 4.2 \times 10^7}$$

$$= \mathbf{1°C \text{ or } 1\ K.}$$

Problem 36:

Calculate the change in the melting point of naphthalene for one atmosphere rise in pressure given that its melting point is 80°C. Latent heat of fusion is 4563 cal/mol and increase in volume on fusion is 18.7 cm³ per mol and I calorie = 4.2 × 10⁷ ergs.

Solution:

We have $$\frac{dP}{dT} = \frac{L}{T(V_2 - V_1)}$$

$$dT = 1 \text{ atmosphere}$$
$$= 76 \times 13.6 \times 980 \text{ dynes/cm}^2$$

$$dT = ?$$

$$L = 4563 \text{ cal/mol}$$
$$= 4563 \times 4.2 \times 10^7 \text{ ergs/mol}$$

$$V_2 - V_1 = 18.7 \text{ cm}^3\text{/mol}$$

$$T = 80 + 273$$
$$= 353 \text{ K}$$

$$dT = \frac{dP.T.(V_2 - V_1)}{L}$$

$$= \frac{76 \times 13.6 \times 980 \times 373 \times 18.7}{4563 \times 4.2 \times 10^7}$$

$$= \mathbf{0.03488\ K.}$$

Therefore, the increase in the melting point of naphthalene with an increase in pressure of one atmosphere

$$= \mathbf{0.03488\ K}$$
$$= \mathbf{0.03488°C.}$$

Problem 37:

Calculate the boiling point of benzene under a pressure of 80 cm of mercury. The normal boiling point is 80°C. Latent heat of vaporization is 389 joules/g, density of vapour at the boiling point is 4 g/litre and that of the liquid 0.9 g/cm³.

Solution:

Here $$dP = 80 - 76$$
$$= 4 \text{ cm of Hg}$$
$$= 4 \times 13.6 \times 980 \text{ dynes/cm}^2$$

$$T = 80°C = 80 + 273$$

$$= 353 \text{ K}$$

$$E = 380 \text{ joules/g}$$

$$= 380 \times 10^7 \text{ ergs/g}$$

$$V_1 = \frac{1}{0.9} = 1.11 \text{ cm}^3$$

$$V_2 = \frac{1{,}000}{4} = 250 \text{ cm}^3$$

$$\frac{dP}{dT} = \frac{L}{T[V_2 - V_1]}$$

$$dT = \frac{dP \times T[V_2 - V_1]}{L}$$

$$dT = \frac{4 \times 13.6 \times 980 \times 373 \times [250 - 1.11]}{380 \times 10^7}$$

$$\mathbf{= 1.233\ K}$$

$$\mathbf{= 1.233°C.}$$

∴ The boiling point of benzene at a pressure of 80 cm of Hg

$$= 80 + 1.233$$

$$\mathbf{= 81.233°C.}$$

Problem 38:

Calculate the change in the boiling point of water when the pressure of steam on its surface is increased from 1 atmosphere to 1.10 atmospheres.

Latent heat of water at 100°C

= 537 cal/g

Volume of one-gram of steam at 100°C

= 1.676 cm³.

Solution:

We have dP = 1.10 − 1

= 0.10 atmosphere

$$dP = 0.1 \times 76 \times 13.6 \times 980 \text{ dynes/cm}^2$$

$$T = 100 + 273$$

$$= 373 \text{ K}$$

$$L = 537 \text{ cals/g}$$

$$= 537 \times 4.2 \times 10^7 \text{ergs/g}$$

$$V_1 = 1.00 \text{ cm}^3$$

$$V_1 = 1.676 \text{ cm}^3$$

$$\frac{dP}{dT} = \frac{L}{T[V_2 - V_1]}$$

$$dT = \frac{dP \times T[V_2 - V_1]}{L}$$

$$dT = \frac{0.1 \times 76 \times 13.6 \times 980 \times 373 \times (1{,}675)}{537 \times 4.2 \times 10^7}$$

$$= \mathbf{2.792 \text{ K}}$$

$$= \mathbf{2.792°C.}$$

Therefore, the increase in the boiling point of water with an increase of 0.1 atmosphere pressure

$$= \mathbf{2.792 \text{ K}}$$

$$= \mathbf{2.792°C.}$$

Problem 39:

Calculate the change in the melting point of ice when it is subjected to a pressure of 100 atmospheres.

Density of ice $= 0.917$ *g/cm³, and*

Latent heat of ice $= 336$ *J/g.*

Solution:

We have $\frac{dP}{dT} = \frac{L}{T(V_2 - V_1)}$

$$dP = 100 - 1$$

$$= 99 \text{ atmospheres}$$

$$dP = 99 \times 76 \times 13.6 \times 980 \text{ dynes/cm}^2$$

$$L = 336 \text{ J/g}$$

$$= 336 \times 10^7 \text{ ergs/g}$$

$$T = 273\text{K}$$

$$(V_2 - V_1) = 1 - \frac{1}{0.917}$$

$$= -\frac{0.083}{0.917}$$

$$= -0.091 \text{ cm}^3$$

$$\therefore \quad dT = \frac{TdP(V_2 - V_1)}{L}$$

$$= \frac{273 \times 99 \times 76 \times 13.6 \times 980 \times (-0.091)}{336 \times 10^7}$$

$$dT = -0.7326 \text{ K}$$

$$= \mathbf{-0.7326°C.}$$

The decrease in the melting point of ice with a pressure of 100 atmospheres

Problem 40:

Calculate the pressure required to lowe melting point of ice by 0°C. (L = 79.6 cal/g, specific volume of water at 0°C = 1.000 cm² specific volume of ice at 0°C = 1.091 cm³ and 1 atmosphere pressure = 1.013 × 10⁶ dynes/cm²).

Solution:

We have $\frac{dP}{dT} = \frac{L}{T(V_2 - V_1)}$

$$dT = -1 \text{ K}$$

$$T = 273 \text{ K}$$

$$V_2 - V_1 = -0.091 \text{ cm}^3$$

$$L = 79.\text{-}6 \text{ cal/g}$$

$$= 79.6 \times 4.18 \times 10^7 \text{ ergs/g}$$

$$dP = \frac{L.dT}{T(V_2 - V_1)}$$

$$dP = \frac{79.6 \times 4.18 \times 10^7 \times 1}{273 \times 0.091} \text{ dynes/cm}^2$$

or $$dP = \frac{79.6 \times 4.18 \times 10^7}{273 \times 0.091 \times 1013 \times 10^6} \text{ atmospheres}$$

$$dP = 135.2 \text{ atmospheres}$$

∴ Pressure required

$$= 135.2 + 1$$

$$= \mathbf{136.2 \text{ atmospheres.}}$$

Problem 41:

Water boils at a temperature of 101°C at a pressure of 787 mm of Hg. 1 gram of water occupies 1,601 cm³ on evaporation. Calculate the latent heat of steam. J = 4.2 × 10⁷ ergs/cal.

Solution:

We have $$\frac{dP}{dT} = \frac{L}{T(V_2 - V_1)}$$

$$dP = 787 - 760$$

$$= 27 \text{ mm of Hg}$$

$$= 2.7 \text{ cm of Hg}$$

$$= 2.7 \times 13.6 \times 980 \text{ dynes/cm}^2$$

$$dT = 1°C = 1K$$

$$T = 373 \text{ K}$$

$$V_2 - V_1 = 1{,}601 - 1 = 1{,}600 \text{ cm}^3$$

$$L = ?$$

$$L = \frac{L\,dP\,(V_2 - V_1)}{dT}$$

$$L = \frac{373 \times 2.7 \times 13.6 \times 980 \times 1{,}600}{1} \text{ ergs/g}$$

or $$L = \frac{373 \times 2.7 \times 13.6 \times 980 \times 1{,}600}{4.2 \times 10^7} \text{ cal/g}$$

$$\mathbf{L = 511.3 \text{ cal/g.}}$$

Problem 42:

When lead is melted at atmospheric pressure, (the melting point is 600 K) the density decreases from 11.01 to 10.66 g/cm³ and the latent heat of fusion is 24.5 J/g. What is the melting point at a pressure of 100 atmospheres?

Solution:

We have $$\frac{dP}{dT} = \frac{L}{T(V_2 - V_1)}$$

$$dP = 99 \text{ atmospheres}$$

$$= 99 \times 76 \times 13.6 \times 980 \text{ dynes/cm}^2$$

$$L = 24.5 \text{ J/g}$$

$$= 24.5 \times 10^7 \text{ ergs/g}$$

$$V_1 = \frac{1}{11.01} \text{ cm}^3$$

$$V_2 = \frac{1}{10.65} \text{ cm}^3$$

$$V_2 - V_1 = \frac{1}{10.65} - \frac{1}{11.01}$$

$$= \left(\frac{0.36}{10.65 \times 10.01}\right) \text{cm}^3$$

$$dT = \frac{TdP\,(V_2 - V_1)}{L}$$

$$dT = \frac{600 \times 99 \times 76 \times 13.6 \times 980 \times 0.36}{10.65 \times 11.01 \times 24.5 \times 10^7}$$

$$\mathbf{dT = 0.7539 \text{ K}}$$

$$\mathbf{= 0.7539°C.}$$

∴ Melting point of lead at 100 atmospheres pressure

$= 600 + 0.7539$

$\mathbf{= 600.7539\ K.}$

Problem 43:

Calculate under what pressure water will boil at 120°C, if the change in specific volume when 1 gram of water is converted into steam is 1,676 cm³.

Latent heat of steam

$= 540$ *cal/g*

$J = 4.2 \times 10^7$ *ergs/cal*

1 atmosphere pressure

$= 10^6$ *dynes/cm²*.

Solution:

Here $\dfrac{dP}{dT} = \dfrac{L}{T(V_2 - V_1)}$

$dT = 120 = 100$

$= 20\ K$

$T = 373\ K$

$(V_2 - V_1) = 1{,}676\ cm^3$

$L = 540\ cal/g$

$= 540 \times 4.2 \times 10^7\ ergs/g$

$dP = ?$

$$dP = \frac{L.dT}{T(V_2 - V_1)}$$

$$dP = \frac{540 \times 4.2 \times 10^7 \times 20}{\cdot 373 \times 1{,}676}\ \text{dynes/cm}^2$$

$$dP = \frac{540 \times 4.2 \times 10^7 \times 20}{373 \times 1{,}676 \times 10^6}\ \text{atmospheres}$$

$dP = 0.7254$ atmosphere

Pressure required $= 1 + 0.7254$

= 1 + 0.7254 atmospheres.

Problem 44:

Calculate the change in entropy when 10 grams of ice at 0°C is converted into water at the same temperature.

Solution:

Heat absorbed by 10 g of ice at 0°C when it is converted into water at 0°C = 10 × 80 = 800 cal

$$\therefore \quad \delta H = 800 \text{ cal}$$

$$T = 0°C = 273 \text{ K}$$

The gain in entropy

$$dS = \frac{\delta H}{T}$$

$$= \frac{800}{273} = \mathbf{2.93\ cal/K}.$$

Problem 45:

Calculate the change in entropy when 5 kg of water at 100°C is converted into steam at the same temperature.

Solution:

Heat absorbed by 5 kg of water at 100°C when it is converted into steam at 100°C

$$= 5000 \times 540$$

$$= 2700000 \text{ cal}$$

$$SB = 2700000 \text{ cal}$$

The gain in entropy

$$dS = \frac{\delta H}{T}$$

$$= \frac{270000}{373} = \mathbf{72\ cal/K.}$$

Problem 46:

Calculate the increase in entropy when 1 gram of ice at –10°C is Converted into steam at 100°C. Specific heat of ice a 0.5, latent heat of ice = 80 cal/g, latent heat of steam = 540 cal/g.

Solution:

(1) Increase in entropy when the temperature of 1 gram of ice increases from –10°C to 0°C

$$dS = \int_{T_1}^{T_2} \frac{\delta H}{T}$$

$$= ms\int_{T_1}^{T_2} \frac{dT}{T}$$

$$= ms \log_e \frac{T_2}{T_1}$$

$$= ms \times 2.3026 \log_e \frac{T_2}{T_1}$$

$$= 1.0.5 \times 2.3026 \log_{10} \frac{273}{263}$$

$$= \mathbf{0.01865\ cal/K.}$$

(2) Increase in entropy when 1 gram of ice at 0°C is converted into water at 0°C.

$$dS = \frac{\delta H}{T}$$

$$= \frac{80}{273} = 0.293 \text{ cal/K}$$

(3) Increase in entropy when the temperature of 1 f of water is raised from 0°C to 100°C.

$$dS = \int_{T_1}^{T_2} \frac{\delta H}{T}$$

$$= ms \times 2.3026 \log_{10} T_2/T_1$$

$$= 1 \times 1 \times 2.3026 \log_{10} 373/273$$

$$= \mathbf{0.312\ cal/K.}$$

(4) Increase in entropy when 1 g water at 100°C is converted into steam at 100°C

$$dS = \frac{\delta H}{T}$$

$$= \frac{540}{373} = 1.447 \text{ cal/K}$$

Total increase in entropy

$$= 0.01865 + 0.293 + 0.312 + 1.447$$

$= \mathbf{2.07065\ cal/K.}$

Problem 47:

One gram molecule of a gas expands isothermally to four times its volume. Calculate the change in its entropy in terns of the gas constant.

Solution:

Work done $= \int_{V_1}^{V_2} PdV$

But $PV = RT$

or $P = \frac{RT}{V}$

$$W = RT \int_{V_1}^{V_2} \frac{dV}{V}$$

$$= RT \log_e \frac{V_2}{V_1}$$

Here $\frac{V_2}{V_1} = 4$

$$W = RT \times 2.3026 \log_{10}(4)$$

Here, W and R are in the units of work

Gain in entropy $= \delta H/T$

$$= \frac{W}{JT} = \frac{RT \times 2.3026 \log_{10} 4}{JT}$$

$$= \mathbf{1.387\ R/J\ cal/K.}$$

Problem 48:

50 grams of water at 0°C if mixed with an equal mass of water at 83°C. Calculate the resultant increase in entropy.

Solution:

(i) $m_1 = 50$ g; $T_1 = 273$ K

$m_1 = 50$ g; $T, = 353$ K

Let the final temperature of the mixture be T K

$$m_1\ 8 \times (T - T_1) = m_2\ s(T_2 - T)$$

$$50 \times 1 \times (T - 273) = 50 \times 1 \times (353 - T)$$

$$\mathbf{T = 310\ K.}$$

(ii) Change in entropy by 50 g of water when its temperature from 273 K to 313 K.

$$= \frac{\delta H}{T} = ms \int_{T_1}^{T} \frac{dT}{T}$$

$$= 50 \times 1 \times \log_e \frac{313}{273}$$

$$= 50 \times 2.3026 \times \log_e \frac{313}{273}$$

$$= \mathbf{+6.829\ cal/K.}$$

Here, the +ve sign indicates gain in entropy.

(iii) Change in entropy by 50 g of water when its temperature falls from 353 K to 313 K

$$= \frac{\delta H}{T} = ms \int_{T_2}^{T} \frac{dT}{T}$$

$$= 50 \times 1 \times \log_e \frac{313}{353}$$

$$= 50 \times 2.3026 \times \log_{10} \frac{313}{353}$$

$$= -6.023\ cal/K.$$

Here, the –ve sign indicates loss in entropy.

Therefore, the total gain in entropy of the system

$$= 6.829 - 6.023$$

$$= \mathbf{0\ 806\ cal/K.}$$

Problem 49:

Calculate the change in entropy when 50 grams of water at 15°C is mixed with 80 grams of water at 40°C. Specific heat of wetter may be assumed to be equal to 2.

Solution:

(i) $m_1 = 50g$

$T_1 = 15 + 273 = 288$ K

$m_2 = 10$ grams

$T_2 = 40 + 273 = 313$ K.

Let the final temperature be T K.

$$m_1 \times 8 \times (T - T_1) = m_2 \times 8 \times (T_2 - T)$$

$$50 \times 1 \times (T - 288) = 80 \times 1 \times (313 - T)$$

$$\mathbf{T = 303\text{-}4\ K.}$$

(ii) Change in entropy when the temperature of 50 g of water rises from 288 K to 303.4 K

$$= \frac{\delta H}{T} = ms \int_{T_1}^{T} \frac{dT}{T}$$

$$= 50 \times 1 \times 2.3026 \times \log_{10} \frac{303.4}{288}$$

$$= \mathbf{+2.602\ cal/K.}$$

(iii) Change in entropy when the temperature of 80 g of water decreases from 313 K to 303.4 K

$$= \frac{\delta H}{T} = ms \int_{T_2}^{T} \frac{dT}{T}$$

$$= 80 \times 1 \times 2.3026 \times \log_{10} \frac{303.4}{313}$$

$$= \mathbf{-2.487\ cal/K.}$$

Therefore, the net change in the entropy of the system

$$= +2.602 - 2.487$$

$$= +0.115 \text{ cal/K}$$

Hence the net increase in the entropy of the system

$$= \mathbf{0.115\ cal/K.}$$

Problem 50:

10 g of steam at 100°C is blown into 90 grams of water at 0°C, contained in a calorimeter of water equivalent 19 grams. The whole of the steam is condensed. Calculate the increase in the entropy of the system.

Solution:

(i) $m_1 = 10$g

$T_1 = 100°C = 373$ K

$m_2 = 90 + 10 = 100g$

$T_2 = 273$ K

Let the final temperature be T K

$10 \times 540 + 10(373 - 7) = 100(T - 273)$

$$T = 331.2 \text{ K.}$$

(ii) Change in entropy when the temperature of water and calorimeter rises from 273 K to 331.2 K

$$= \frac{\delta H}{T} = ms \int_{T_2}^{T} \frac{dT}{T}$$

$$= 100 \int_{273}^{331.2} \frac{dT}{T}$$

$$= 100 \times 2.3026 \times \log_{10}\left(\frac{331.2}{273}\right)$$

$$= +19.32 \text{ cal/K.}$$

(iii) Change in entropy when 10 grams of steam at 373 K is condensed to water at 373 K

$$= \left(\frac{\delta H}{T}\right) = -\frac{10 \times 540}{273}$$

$$= -14.47 \text{cal/K.}$$

(–ve sign indicates decrease in entropy).

(iv) Change in entropy when 10 grams of water at 373 K is cooled to water at 331.2 K

$$= \frac{\delta H}{T} = ms \int_{T_2}^{T} \frac{dT}{T}$$

$$= 10 \times 2.3026 \log_{10}\left(\frac{331.2}{373}\right)$$

$$= -1.188 \text{ cals/K}$$

Net change in entropy

$$= 19.32 - 14.47 - 1.188$$

$$= +3.662 \text{ cal/K.}$$

Hence the net increase in the entropy of the system

$$= 3.662 \text{ cal/K.}$$

4

THERMAL CONDUCTORS AND STATISTICAL POSTULATES

INTRODUCTION

Thermal conductivity of poor conductors viz., rubber, glass, ebonite, wood, cork etc. The specimen is taken in the form of two thin discs D_1, and D_2 about 10 cm in diameter and 2 to 3 mm thick. The disc A is pressed between two copper plates C_1 and C_2 and D_2 is pressed between the copper plates C_3 an d C_4 Discs C_1 and C_3 ensure normal flow of heat through the experimental plates D_1 and D_2. H is a heater coil and T_1, T_2, T_3 and T_4 are four thermocouples used to measure temperatures. The surfaces of D_1 and D_2 are coated with glycerine so that these surfaces make good thermal contact with the copper plates. A steady current I is passed through the heater coil H. The potential difference across the heater coil is E.

After the steady, state is reached the temperatures of the thermometers T_1, T_2, T_3 and T_4 are noted. Let the temperatures be θ_1, θ_2, θ_3 and θ_4 respectively. In the steady state, heat generated in the heater coil is lost from the surface of C_2 and C_4, and heat lost from the rims of C_1, C_2, D_1 and D_2 is negligible due to the small thickness of the plates.

Let d_1 be the thickness of the disc D_1 and d_2 the thickness of the disc D_2.

Heat produced by heater coil in one second

$$= \frac{EI}{4.2} \text{ calories.} \qquad \text{...(i)}$$

Heat passing through D_1 and D_2, in one second

$$= \frac{KA(\theta_1 - \theta_2)}{d_1} + \frac{KA(\theta_3 - \theta_4)}{d_2} \qquad \text{...(ii)}$$

Equating (i) and (ii)

$$\frac{EI}{4.2} = \frac{KA(\theta_1 - \theta_2)}{d_1} + \frac{KA(\theta_3 - \theta_4)}{d_2} \qquad ...(iii)$$

The value of K is calculated from equation (iii).

BAD CONDUCTORS

The apparatus consists of a cylindrical steam chamber A, the specimen disc D and brass or copper block C. The whole apparatus is suspended from the stand. T_1 and T_2 are the thermometers used to determine the temperature after the steady state reached.

Steam is passed through the chamber and the readings of the thermometers T_1 and T_2 are noted after the steady state is reached. The heat passing through D in one second is equal to the heat radiated by the exposed surface of C in one second.

$$\frac{KA(\theta_1 - \theta_2)}{d} = ms\frac{d\theta}{dt}\left[\frac{A+S}{2A+S}\right] \qquad ...(i)$$

Here $\left[\frac{A+S}{2A+S}\right]$ is the fraction of the total area exposed to the surroundings.

Here A is the area of cross-section of D and C. S is the area of the curved surface C, $\frac{d\theta}{dt}$ is the rate of cooling at temperature θ_2, m is the mass and s is the specific heat of C.

To find $\frac{d\theta}{dt}$, the disc D is removed and C makes contact with the steam chamber. C is removed when its temperature is about 10°C higher than θ_2. It is placed over two knife edges and its temperature is observed after equal intervals of time (say one minute). A graph is drawn between temperature and time. From the graph, the value of $\frac{d\theta}{dt}$ at temperature θ_2 is found.

From equation (i), K can be calculated.

LEE'S METHOD FOR LIQUIDS

To determine the thermal conductivity of a liquid, convection current should be determined. To overcome this difficulty, heat should flow from the upper surface to the lower surface of the liquid.

Lee designed an apparatus C_1, C_2, C_3 and C_4 are the copper plates of uniform thickness. H is the heater coil. A glass plate is pressed between C_2 and C_3. The thermocouples T_1, T_2, T_3 and T_4 are used to determine the temperatures. The liquid is contained in a cylindrical tube having an ebonite ring.

A steady current of I amperes is passed through the heater coil H. After the steady state is reached, the temperatures shown by the thermocouples are noted. Let these temperatures be θ_1, θ_2, $\theta_2\theta_3$ and θ_4.

Area of cross-section of the glass plate = A

Thickness of the glass plate = d

Area of cross-section of the liquid surface = A_1

Area of cross-section of the ebonite surface = A_2

Thickness of the liquid = d_1

Thickness of the ebonite ring = d_1

Thermal conductivity of glass = K

Thermal conductivity of liquid = K_1

Thermal conductivity of ebonite = K_2

Heat flowing per second through the glass plate

$$= \frac{KA(\theta_1 - \theta_2)}{d} \quad \text{...(i)}$$

Heat flowing per second through the liquid and the ebonite ring

$$= \frac{K_1A_1(\theta_3 - \theta_4)}{d_1} + \frac{K_2A_2(\theta_3 - \theta_4)}{d_1} \quad \text{...(ii)}$$

Equating (i) and (ii)

$$\therefore \quad = \frac{KA(\theta_1 - \theta_2)}{d} = \frac{K_1A_1(\theta_3 - \theta_4)}{d_1} + \frac{K_2A_2(\theta_3 - \theta_4)}{d_1} \quad \text{...(iii)}$$

The value of K, for the liquid can be calculated from equation (iii). However, there is small error due to the heat lost by the sides of C_1, C_2, C_3 C_4, the glass plate and the ebonite ring.

SPHERICAL SHELL METHOD

Consider two thin spherical shells A and B of radii r_1 and r_2. The specimen is contained between these two shells A and B. A heating element is placed at the centre of the shells. Heat is conducted through the specimen from the inner to the outer shell. Let the temperatures of the inner and the outer shells be θ_1 and $\theta_1\theta_2$, after the steady state is reached. Consider an imaginary shell of radius r and thickness dr having temperatures θ and $\theta + d\theta$ on its inner and the outer surfaces respectively. The quantity of heat conducted per second through this shell,

$$Q = -KA\frac{d\theta}{dr} \qquad ...(i)$$

Here $$A = 4\pi r^2$$

$\therefore$ $$Q = -K.4\pi r^2\frac{d\theta}{dr}$$

$$\frac{dr}{r^2} = -\frac{4\pi K}{Q}d\theta$$

Integrating,

$$\int_{r_1}^{r_2}\frac{dr}{r^2} = -\frac{4\pi K}{Q}\int_{\theta_2}^{\theta_2}d\theta$$

$$\frac{1}{r_1} = \frac{1}{r_2} = \frac{4\pi K(\theta_1-\theta_2)}{Q}$$

$$K = \frac{Q(r_2-r_1)}{4\pi r_1 r_2(\theta_1-\theta_2)}$$

K is calculated from equation (ii).

WIEDEMANN-FRANZ LAW

Wiedemann and Franz, based on experimental results, put forward a law relating to the thermal conductivity and electrical conductivity of metals. According to this law t.

The ratio of the thermal and electrical conductivities is the same for all metals at the same temperature. Moreover, the ratio is directly proportional to the absolute temperature of the metal.

Let K and a be thermal and electrical conductivities of a metal at a temperature T degrees absolute,

$$\frac{K}{\sigma} \propto T$$

or $$\frac{K}{\sigma T} = \text{constant for all metals.}$$

This law holds good for a large number of metals between +100°C and –100°C. At low temperatures the ratio K/σ decreases and the value tends to be zero at absolute zero. As the temperature of the metal is decreased, the thermal and electrical conductivities of the metal increase. But the increase in the electrical conductivity is higher and its value tends to infinity at absolute zero.

This corresponds to the super-conducting state of the metal.

Drude explained that thermal and electrical conductivities are due to free electrons in metals. He derived the expression

$$\frac{K}{\sigma T} = \frac{3R^2}{JN^2e^2} = \text{constant},$$

where R is the universal gas constant, N is the number of molecules in one gram molecule J is the Joule's mechanical equivalent of heat and e is the charge on the electron.

Values of $\frac{K}{\sigma T} \times 10^6$.

Metal	–100°C	–50°C	°C	18°C
Aluminium	1.81	1.98	2.09	2.13
Copper	2.17	2.26	2.30	2.32
Iron	2.98	2.93	2.97	2.99
Lead	2.54	2.52	2.53	2.51
Silver	2.29	2.36	2.33	2.33
Zinc	2.39	2.40	2.45	2.43

CONDUCTION OF HEAT

Metals are good conductors of heat and wood, felt, brick, glass, granite, cotton, wool, cork, ebonite, rubber are bad conductors of heat.

(1) Sauce pans, hot water buckets, kettles and other utensils are made of metal. They are provided with wooden or ebonite handles so that heat from the utensil is not conducted to the hand.

(2) Ice box has a double wall made of tin or iron. The space in between the two walls is packed with cork or felt. This is done because cork and felt are poor conductors of heat and prevent the flow of the outside heat into the box.

(3) Thick brick walls are used in the construction of a cold storage. Brick is a bad conductor of heat and does not allow outside heat to flow inside the cold storage.

(4) Woollen clothes have fine pores filled with air. Air and wool are bad conductors of heat. Heat from the body does not flow outside to the atmosphere. Therefore woollen clothes keep the body warm in winter.

(5) Two shirts keep the body warmer than a single shirt of the same material and double the thickness. Between the two shirts a fine layer of air acts as a bad conductor and does not allow the heat from the body to flow out to the surroundings.

(6) Quilts and bed clothings filled with cotton are used in winter. The air layers in the pores of the cotton are bad conductors of heat. Therefore the flow of heat to outside a prevented.

(7) A steel blade appears colder than a wooden handle in winter. Steel is a good conductor of heat. As soon as a person touches the blade, heat flows from the hand to the blade. Therefore it appears colder. Since wood is a bad conductor of heat, heat does not flow from the hand to the handle.

(8) In cold countries, the windows are provided with double doors. Air in between the two doors forms a non-conducting layer and does not allow heat to flow from inside to outside.

In hot countries also, double door windows are used. Heat does not flow from outside to inside because air forms a non-conducting layer in between the two doors.

(9) When a stopper fitted tightly to the bottle is to be removed the neck is gently heated. Glass is a poor conductor of heat. Neck expands but heat is not conducted to the stopper. The stopper can be removed easily.

ACCRETION OF ICE ON PONDS

Consider a layer of ice x cm thick on the surface of a pond. Let the temperature of air over the surface of ice be θ°C and that of water below ice 0°C. Suppose a thickness dx of ice is formed in time dt.

Mass of ice formed

$$= A.dx.\rho$$

Here A is the area of the pond, ρ the density of ice and L the latent heat of fusion of ice.

Heat lost by water

$$= A.dx.\rho.L \text{ calories} \qquad ...(i)$$

This heat is conducted across a layer of ice of thickness x upwards.

Heat conducted $\quad KA \frac{\theta}{x} dt$ calories $\qquad ...(ii)$

Equating (i) and (ii)

$$KA \frac{\theta}{x} dt = Adx.\rho L$$

$$\frac{dx}{dt} = \frac{K\theta}{\rho Lx} \qquad ...(iii)$$

$\frac{dx}{dt}$ represents the rate of growth of the thickness of ice.

$$dt = \left(\frac{\rho L}{K\theta}\right) xdx$$

Total time taken by the layer of ice to increase in thickness by x

$$\int dt = \frac{\rho L}{K.\theta} \int x.dx$$

$$t = \frac{\rho L}{K.\theta} \frac{x^2}{2} + \text{constant}$$

When $\quad t = 0, x = 0$

Hence the constant is zero.

$$t = \frac{\rho L}{2K\theta} x^2 \qquad ...(iv)$$

APPLICATION OF CONVECTION

1. Ventilation

Rooms are provided with ventilators near the ceiling. Air in the room gets warmer due to respiration of people in the room. Warm air containing more of CO_2 and water vapour has less density and moves upwards. Fresh air from outside enters the room through the doors and windows. The impure air moves, outside through die ventilators. The phenomenon is continuous.

2. Chimneys

In a kitchen, chimney is provided. Hot air moves up the chimney and fresh air enters the fire place from outside. This provides enough oxygen for the continuous combustion of the fuel Similarly in off banns, oil stoves, principle of a chimney is used. When the wick is lighted, air and kerosene oil burn above the wick. Heated air moves up and fresh air from the holes provided at the bottom of the chimney enters the inside. The phenomenon is continuous and the convection currents set up help in supplying enough oxygen for combustion. If holes are not provided, the flame gets extinguished quickly.

In a factory, a high chimney is provided. The rate of circulation of air depends on the difference in pressure between the two ends of the chimney. Higher the chimney, greater is the rate of circulation of air. Due to this reason, sufficient supply of oxygen is provided and the complete combustion of the fuel takes place.

3. Trade Winds

The surface of the earth gets heated more at the equator than at me poles. Warm air at the equator moves up and cold air from the poles moves towards the equator. In the northern hemisphere, if is coming from the north and due to the rotation of the earth from west to cast, the wind appears to come from north-east. In the southern hemisphere, the wind appears to be from south-west. These winds are called trade winds because they were used by traders for sailing their vessels in ancient days. Monsoons are also based on the principle of convection.

4. Ocean Currents

Water in the ocean at the equator gets heated more than at the polo. Heated water at the equator expands and the level of water rises near the equator. This hot water flows towards the poles. This current is called hot current or gulf streams. The cold water from the poles moves towards the equator beneath the ocean and this current is called the under current.

5. Land and Sea Breezes

Near the sea, water becomes less warm than the land during day time. The heated air on the land surface moves up. Air from the surface of the sea moves towards the sea shore. This is called sea breeze. Convection currents are set up During night when the land becomes cooler than water, the air over the surface of water is warmer and moves upward. Air from the land moves towards the sea.This is called land breeze.

HEATING SYSTEM

In cold countries, the temperature in Winter falls below 0°C. The rooms of a building are kept warm by a central heating system based on the principle of convection. Water is heated in the boiler at the basement of the building. The rooms are fitted with pipes having radiators fixed on the walls. Hot water from the boiler rises up. It passes through the radiators of different rooms. Radiators get heated and radiate heat to the room. After losing heat to the radiator, water becomes cold and returns back to the boiler. Hot water also reaches the cold water tank at the top of the building. The hot water flow a from the boiler to the cold water tank and the cold water flows from the tank to the boiler. Convection currents are set up and the building is kept warm continuously at a constant temperature.

CHANGE OF PRESSURE WITH HEIGHT

The pressure of the atmosphere decreases with increase in height from sea level. The density of air is not the same at all levels. If the density of air were the same at all levels, the atmospheric pressure at sea level would correspond to a vertical column of 8 km of air only. But it has been found that air exists even at a height of 200km.

Consider that the pressure of air at a height h = P. For an increase in height dh, the decrease in pressure is dP.

$$dP = -(dh)\rho g \qquad \text{...(i)}$$

[–ve sign shows that the pressure decreases with increase in height]

Let M be the molecular weight of the gas and V its volume.

Then, $$\rho = \frac{M}{V}$$

and $$dP = -(dh)\frac{M}{V}g$$

But for a perfect dry air

$$PV = RT$$

or $$V = \frac{RT}{P}$$

$\therefore$ $$dP = -(dh)\frac{M.Pg}{RT}$$

$$\frac{dP}{P} = -\frac{Mg}{RT}(dh) \qquad ...(ii)$$

Integrating

$$\log P = -\frac{Mg}{RT}h + C \quad \text{[where C is a constant]}$$

When $h = 0,$

$$P = P_0$$

$\therefore$ $$C = \log P_0$$

and $$\log P = -\frac{Mg}{RT}h + \log P_0$$

$$\log_e \frac{P}{P_0} = -\frac{Mg}{RT}h \qquad ...(iii)$$

$$P = P_0 e^{-\frac{Mg}{RT}h} \qquad ...(iv)$$

Equation (ii) will be true only under isothermal conditions. But air temperature decreases uniformly with height. Assuming that the lapse rate is α in the lower portion of the atmosphere (troposphere),

$$T = T_e - \alpha h$$

where T_0 is the temperature of the ground.

Substituting this value in equation (ii)

$\therefore$ $$\frac{dP}{P} = -\frac{Mg}{R}\frac{dh}{(T_0 - \alpha h)}$$

$$\log P = \frac{Mg}{R\alpha}\log(T - \alpha h) + K \qquad ...(v)$$

$$h = 0, P = P_0$$

$\therefore$ $$\log P_0 = \frac{Mg}{R\alpha}\log T_0 + K$$

or $$K = \log P_0 - \frac{Mg}{R\alpha}\log T_0$$

Substituting this value of K in equation (v)

$$\log P = \frac{Mg}{R\alpha} \log (T_0 - \alpha h) + \log P_0 - \frac{Mg}{R\alpha} \log T_0$$

$$\log \frac{P}{P_0} = \frac{Mg}{R\alpha} \log\left(\frac{T_0 - \alpha h}{T_0}\right) \quad \text{...(vi)}$$

$$P = P_0 e^{\frac{Mg}{R\alpha} \log\left(\frac{T_0 - \alpha h}{T_0}\right)} \quad \text{...(vii)}$$

EQUILIBRIUM OF THE ATMOSPHERE

The atmospheric temperature decreases with altitude upto a height of about 20 km. The rate of fall of temperature is about 5oC per km height and is known as the lapse rate. The atmosphere is divided into two regions.

1. *Troposphere or the convective zone* : It is the region in which the temperature falls with increase in height from the ground.
2. *Stratosphere or advective zone*: It is the region in which the decrease in temperature does not take place with increase in height.

The layer that separates these two regions is known as *tropopause*. The height of tropopause is different at different places on the earth. It is about 14 km at the equator and about 10 km at the poles.

The fall in temperature with altitude in the troposphere is due to convection. When the heat radiations from the sun fall on the surface of the earth, it gets heated. The atmospheric air surrounding the earth gets heated by conduction and radiation. The heated air moves up and convection currents are set up. The heated air that moves up from the regions of higher to lower pressure is adiabatically cooled. Similarly when the colder air moves down from lower to higher pressure regions it is adiabatically heated. Thus, a *convective* equilibrium is set up and there is a gradual fall of temperature with increase in height. The earth gets heated up due to the heat radiations from the sun. The layers of the atmosphere close to the ground get heated up and convection currents start in the atmosphere. As the warm air rises up into the region of low pressure, it *expands adiabatically*. For an atmosphere, in convective equilibrium, there is a vertical fall of temperature from the earth's surface. The *dry adiabatic lapse rate* is calculated as follows:

The relation between-pressure and temperature for an adiabatic process is

$$\frac{P^{\gamma-1}}{T^{\gamma}} = \text{const.}$$

or $$P^{\gamma-1}\,T^{-\gamma} = \text{const.}$$

Differentiating

$$(\gamma-1)\,P^{\gamma-2}dP.T^{-\gamma} + P^{\gamma-1}(-\gamma)T^{-\gamma-1}dT = 0$$

Dividing by $P^{\gamma-1}T^{-\gamma}$

$$(\gamma-1)\frac{dP}{P} + (-\gamma)\frac{dT}{T} = 0$$

$$\frac{dP}{P} = \frac{\gamma}{\gamma-1}\frac{dT}{T} \quad \text{...(i)}$$

For an increase in height dh, the decrease in pressure is dP

$\therefore$ $$dP = -\rho g\,dh = -\frac{M}{V}g\,dh$$

But $$PV = RT$$

or $$V = \frac{RT}{P}$$

$\therefore$ $$dP = -\frac{MP}{RT}g\,dh$$

$$\frac{dP}{P} = -\frac{Mg}{RT}dh \quad \text{...(ii)}$$

Equating (i) and (ii)

$$\left(\frac{\gamma}{\gamma-1}\right)\frac{dT}{T} = -\frac{Mg}{RT}dh$$

$$\frac{dT}{dh} = -\frac{Mg}{R}\left(\frac{\gamma-1}{\gamma}\right) \quad \text{...(iii)}$$

This equation holds good only for perfect dry air. The lapse rate $\frac{dT}{dh}$ can be calculated from equation (iii).

Substituting the values,

$$M = 29 \text{ g}$$

$$g = 981 \text{ cm/s}^2$$

$$R = 8.31 \times 10^7 \text{ ergs per gram mol per K,}$$

and $\gamma = 1.4$

$$\frac{dT}{dh} = -\frac{2.9 \times 981}{8.31 \times 10^7}\left(\frac{1.40-1}{1.40}\right)$$

$$= 10^{-4} \text{ K per cm}$$

$$= 10^{-4}\ ^\circ\text{C per cm.}$$

Thus, the lapse rate for a height of 1 km = $10^{-4} \times 10^5$ = 10°C. But it has been found-that the average lapse rate is lower than this value. Under average conditions the lapse rate has been found to be between 5°C and 6-5°C per km.

However, at night, when the earth is radiating heat to the atmosphere, the layers of air nearer the surface may cool to a temperature lower than that of the upper layers. Then a negative lapse rate is set up.

DIFFERENTIAL AIR THERMOSCOPE

A differential air thermoscope consists of two glass bulbs A and B connected by a narrow glass tubing. The U-tube contains sulphuric acid. The bulbs A and B contain air and when they are at the same temperature, there is no difference in level of the liquid in the two limbs.

The bulb A is coated with lamp black so that it can completely absorb the whole of the heat radiations falling on it. When A is exposed to heat radiations, it absorbs heat energy, Air inside the bulb A expands and there is difference in level of the liquid in the two limbs. This thermoscope is very sensitive and can detect heat radiations of even feeble intensity (*i.e.*, from a candle at a large distance).

Thermopile

The thermopile originally designed by *Melloni* consisted of a number of thermocouples joined in series. The thermocouples are made of bismuth and antimony. The junctions on one side are coated with lamp black and are exposed to heat radiations The junctions on the other side are kept cold. The two ends of the thermopile are connected to a sensitive galvanometer.

The deflection in the galvanometer measures the intensity of heat radiations and the scale can be calibrated.

The improved form of a thermopile, which is commonly used, consists of a large number of thermopiles. The bars of antimony and bismuth are arranged in the form of a cube. The junctions are soldered but the bars are insulated from each other with mica strips. All the junctions on one side are coated with lamp-black. The galvanometer is connected to the points A and B. When not in use, the hot face is covered with a metallic cup. Such a thermopile can detect heat radiations from a candle at a distance of 100 metres. When the thermo-electric current produced in the thermopile flows through the galvanometer, there is a deflection in the galvanometer. The galvanometer scale can be calibrated.

The *linear thermopile* is used in research for the study of infrared radiations. It consists of only a few pairs of thermocouples connected to a sensitive galvanometer. The hot junctions are arranged in a line.

The thermopiles can be made sensitive by mounting the thermopile in vacuum and using very short connecting wires. The chief drawback of a thermopile consists in the comparative slowness of its indication and also the long time it takes to return to its zero reading.

Properties of Thermal Radiation

Heat radiations have properties similar to light radiations. They are electromagnetic waves of longer wavelength. The range of wavelengths of heat radiations is from 3 cm to 8×10^{-5} cm.

(1) Heat radiations travel in straight lines

Take three similar cardboards having holes at the same height. Place a red hot ball in front of the hole of the first cardboard. Bring the blackened bulb of the differential air thermoscope near the hole of the third cardboard. When the hot ball, the three holes and the bulb of the thermoscope are in the same straight line, maximum difference in level is noted. When one of the cardboards is displaced, the thermoscope is not affected. This shows that heat radiations travel in straight lines.

(2) Heat radiations travel with the velocity of light

When the sun rises, heat and light are received simultaneously on the earth. During the solar eclipse heat and light are cut off at the same moment. When the eclipse is over, heat and light reach the earth simultaneously. It shows that the heat radiations travel with the velocity of light equal to 3×10^8 m/s.

(3) Heat radiations obey inverse square law

Take a red hot ball. Place the thermopile at a fixed distance from the ball. Note the reading in the galvanometer. Now, place the thermopile at double the distance from the ball. It is found that the deflection of the galvanometer is one-fourth of the deflection in the first case. It shows that heat radiations obey inverse square law.

(4) Heat radiations obey the laws of reflection

Take two tubes A and B hinged at C. Keep the metal ball at the mouth of A. Adjust the position of B to get the maximum difference in level in the thermoscope. It is found that the difference in level is maximum when the two tubes are equally inclined to the normal at C. It means $\angle i = \angle r$.

Take a concave metallic reflecting surface. Direct it towards the sun. A small piece of paper placed at the focus of the mirror catches fire. All the heat radiations incident on the mirror after reflection converge at its focus. Room heaters are provided with concave or elliptical reflectors. The heater element is kept at the focus of the mirrors.

(5) Heat radiations obey the laws of refraction

Take a convex lens and keep a small paper at its focus. When the lens is directed towards the sun's rays, the paper catches fire. The parallel heat radiations from the sun after refraction from the lens converge at its focus.

(6) Heat radiations get diffused when they are incident on a rough and unpolished surface.

(7) Heat radiations can travel through vacuum.

(8) Heat radiations do not affect the medium through which they pass. When sun's rays are allowed to pass through a convex lens, they converge at the focus of the lens. The lens does not get heated but a paper at the focus catches fire.

SOME APPLICATIONS OF HEAT RADIATIONS

(1) White clothes are preferred in summer and dark coloured clothes in winter.

When heat radiations fall on white clothes, they are reflected back. No heat is absorbed by the clothes and a person does not get heat from outside in summer. Dark clothes in winter will absorb the heat radiations falling on them and keep the body warm.

(2) Cooking utensils are blackened at the bottom and polished on the upper surface. Black surface will absorb the whole of the heat from the furnace and the upper surface will not allow the heat inside the utensil to flow out.

(3) Hot water pipes and radiators used in rooms are painted black so that they can radiate maximum amount of heat to the room. The same pipes outside the rooms are painted white so that they do not lose heat to the surroundings.

(4) The thermocouple junction exposed to heat is blackened to absorb maximum quantity of heat.

(5) Polished reflectors are used in electric heaters to reflect maximum heat in the room.

BLACK BODY

A perfectly black body is one which absorbs all the heat radiations (corresponding to all wavelengths) incident on it. When such a body is placed inside an isothermal enclosure, it will emit the full radiation of the enclosure after it is in equilibrium with the enclosure. These radiations are independent of the nature of the substance. Such heat radiations in a uniform temperature enclosure are known as *black body* radiations. Also the black body completely absorbs heat radiations of all wavelengths. Thus the black body also emits completely the radiations of all wavelengths at that temperature.

In practice, a perfectly black body is not available. A body showing close approximation to a perfectly black body can be constructed.

A hollow copper sphere is taken and is coated with lamp black on its inner surface. A fine hole is made and a pointed projection is made just in front of the hole. When the radiations enter the hole, they suffer multiple reflections and are completely absorbed. This body acts as a black body absorber. When this body is placed in a bath at a fixed temperature, the heat radiations come out of the hole. The hole acts as a black body radiator. It should be remembered that only the hole and not the walls of the body, acts as the black body radiator. Wien also constructed a black body in the form of a cylinder. This black body is commonly used now a days.

It consists of a hollow metallic cylinder and fitted with heating coils wound around it. The inner surface of the cylinder is coated with lamp black. The cylinder is placed in concentric porcelain tubes (Fig. 3.50). The temperature is measured with the help of a thermocouple arrangement. Heat radiations emerge out of the hole. The radiations from the inner chamber can be limited with the help of diaphragms provided on the inner side. This hole will act as a black body radiator.

ABSORPTIVE POWER

It is the ratio of the amount of heat absorbed in a given time by the surface to the amount of heat incident on the surface in the same time.

EMISSIVE POWER

It is the ratio of the amount of heat radiations emitted by unit area of a surface in one second to the amount of heat radiated by a perfectly black body per unit area in one second under identical conditions.

The law states that the ratio of the emissive power to the absorptive power for the radiations of a particular wavelength and at a particular temperature is constant for all bodies. Moreover, this ratio is equal to the emissive power of a perfectly black body.

Suppose e_λ and a_λ are the emissive and absorptive powers of a body. Let Q be the quantity of heat radiations incident on the surface in one second, Q_1 is the quantity of heat adsorbed by the surface and Q_2 is the quantity of heat reflected by it.

As the body radiates heat, at all temperatures, the heat emitted by it in one second $= e_\lambda d\lambda$

Total heat falling on the surface $= Q$ ...(i)

Total heat given out by the surface

$$= (Q - Q_1) + e_\lambda \, d\lambda \qquad ...(ii)$$

For equilibrium, (i) and (ii) are equal

$$\therefore \quad (Q - Q_1) + e_\lambda d\lambda = Q$$

But

$$a_\lambda = \frac{Q_1}{Q}$$

$$Q_1 = a_\lambda Q$$

$$Q - a_\lambda Q + e_\lambda\, d\lambda = Q.$$

or $$\frac{e_\lambda}{a_\lambda} = \frac{Q}{d\lambda} = \text{const.} \qquad \text{...(iii)}$$

For a perfectly black body, emissive power = E_λ and absorptive power $a_\lambda = 1$.

$$\therefore \qquad \frac{E_\lambda}{1} = \frac{Q}{d\lambda} \qquad \text{...(iv)}$$

From equations (iii) and (iv)

$$\frac{e_\lambda}{a_\lambda} = E_\lambda = \text{const} \qquad \text{...(v)}$$

Result: If (iii) is large, a_λ is also large. But

$$r_\lambda = \frac{Q - Q_1}{Q}$$

$$= 1 - a_\lambda.$$

Therefore r_λ is small. Here r_λ is the reflecting power of the surface. It means good emitters are good absorbers but bad reflectors. Dull black surfaces are good emitters and good absorbers but bad reflectors.

If e_λ is small, a_λ is also small and r_λ is large. It means bad emitters are bad absorbers but good reflectors. Highly polished surfaces are bad emitters and bad absorbers but good reflectors.

These results are verified by Ritchie's experiment.

RITCHIE'S EXPERIMENT

It consists of two cylindrical vessels A and B connected to the glass tubing XY. One set of faces of A and B a blackened while the other set of faces is polished. C is another cylindrical vessel containing hot water and fixed on a stand between A and B so that the three cylinders are coaxial. One face of C is blackened while the other is polished. XY serves as a differential air thermoscope.

Initially, the blackened face of C is towards the polished face of A and the polished face of C is towards the blackened face of B. The level of the liquid in the limbs X and Y will be the same. Now, turn C so that the polished face of C is towards the polished face of A. The quantity of heat emitted by the polished face of C is very small. Also the quantity of heat absorbed by the polished face of A is small. Therefore the rise in temperature of air inside A is very small. On the other hand the quantity of heat emitted by the blackened face of C is large and also the quantity of heat absorbed by the blackened face of B is large. Therefore, rise in temperature of air inside B is higher than that in A. Consequently, the level of liquid in the limb Y is lower than that in X.

Hence, good absorbers are good emitters and bad absorbers are bad emitters,

PREVOST THEORY OF HEAT EXCHANGE

According to this theory, every body radiates heat continuously at all temperatures. The quantity of heat radiated per unit area of the surface in unit time is dependent on the temperature of the body and not on the surroundings. Suppose a body A is at a higher temperature than the surroundings. It will radiate more quantity of heat to the surroundings and it will absorb less heat from the surroundings. Consequently, the temperature of the body falls. When the temperature of the body becomes equal to that of the surroundings, it is in thermal equilibrium with the surroundings. The amount of heat radiated is equal to the amount of heat absorbed by it. Similarly, when a body is at a temperature lower than that of the surroundings, it will radiate less heat and absorb more heat.

Thus, all bodies at all temperatures are in a state of dynamic thermal equilibrium when they are at the same temperature. Moreover, the amount of heat radiated from a body decreases with the fall in temperature of the body. When the body is at the absolute zero temperature (–273°C), it will not radiate any heat energy because the kinetic energy of the molecules is zero at the absolute zero temperature.

CYLINDRICAL FLOW OF HEAT

Consider a cylindrical tube of length l, inner radius r_1 and outer radius r_2. After the steady state is reached, the temperature on the inner surface is θ_1 and on the outer surface it is θ_2. Here $\theta_1 > \theta_2$. Heat is conducted radially across the wall of the tube. Consider an element of thickness dr and length l at a distance r from the axis.

The quantity of heat flowing per second across the element

$$Q = -KA\frac{d\theta}{dr} \qquad \text{...(i)}$$

But $\qquad A = 2\pi r l$

$$\therefore \qquad Q = -K.2\pi r l \frac{d\theta}{dr}$$

$$\frac{dr}{r^2} = -\frac{4\pi K}{Q} d\theta$$

$$Q.\frac{dr}{r} = -2\pi r l d\theta$$

Q is constant after the steady state is reached.

Integrating equation (i)

$$Q\int_{r_1}^{r_2} \frac{dr}{r} = -2\pi K l \int_{\theta_1}^{\theta_2} d\theta$$

$$Q\left[\log_e \frac{r_2}{r_1}\right] = -2\pi K l\,[\theta_2 - \theta_1]$$

$$Q\cdot\log_e \frac{r_2}{r_1} = -2\pi K l\,[\theta_1 - \theta_2]$$

$$K = \frac{Q \log_e \frac{r_2}{r_1}}{2\pi l\,(\theta_1 - \theta_2)}$$

$$K = \frac{Q \times 2.3026 \times \log_{10} \frac{r_2}{r_1}}{2\pi l\,(\theta_1 - \theta_2)}$$

THERMAL CONDUCTIVITY OF RUBBER

The coefficient of thermal conductivity of a rubber tubing can be determined in the laboratory applying the principle of cylindrical flow of heat. A known quantity of water is taken in a calorimeter C. A. rubber tubing whose, inner and outer radii are r_1 and r_2, is taken and a known length (say 50 cm) of it is immersed in water. The initial temperature of water is noted. Let it be θ_3. Steam is passed through the rubber tubing for a known time t seconds (say 900 seconds). Let the final temperature of water be θ_4, (after applying radiation correction) and temperature of steam be θ_1. The average temperature on the outer surface of the rubber tubing

$$= \theta_2 = \frac{\theta_4 - \theta_3}{2}$$

Suppose, mass of water = m

Water equivalent of the calorimeter = w

Rise in temperature = $(\theta_4 - \theta_3)$

Heat gained by water = $(m + w)(\theta_4 - \theta_3)$

Quantity of heat flowing per second

$$Q = \frac{(m+w)(\theta_4 - \theta_3)}{t}$$

But $K = \dfrac{Q \times 2.3026 \times \log_{10}\left(\dfrac{r_2}{r_1}\right)}{2\pi l\,[\theta_1 - \theta_2]}$

Substituting the values of θ_2 and Q

$$K = \frac{(m+w)(\theta_4 - \theta_3) \times 2.3026 \times \log_{10}\left(\dfrac{r_2}{r_1}\right)}{2\pi l\left(\theta_1 - \dfrac{\theta_3 - \theta_4}{2}\right)t}$$

Thus, K for rubber can be calculated.

THERMAL CONDUCTIVITY OF GLASS

The apparatus is kept in the slanting position and water is allowed to flow inside the glass tube from lower to the upper end. The thermometers T_1 and T_2 are used to determine the temperature of incoming and outgoing water. The rate of flow of water can be adjusted with the help of a pinch cock. Steam is passed inside the outer jacket and the rate of flow of water is adjusted such that the two thermometers show sufficient difference of temperature (say 5 to 10°C). When the two thermometers show constant temperature, water flowing through the glass tube for a known time t seconds (say 900 seconds) is collected in a beaker. The spiral wire inside the glass tube helps in making good contact of water with the wall of the tube.

Suppose,

Length of the tube inside the jacket = l

Temperature of steam = θ_1

Temperature of incoming water = θ_3

Temperature of outgoing water = θ_4

Mass of water collected = m

Time taken = t seconds

Quantity of heat gained by water in t seconds

$$= m\,(\theta_4 - \theta_3)$$

Quantity of heat that flows in one second across the wall of the tube,

$$Q = \frac{m(\theta_4 - \theta_3)}{t}$$

Average temperature of the inside of the glass tube

$$= \theta_2 = \frac{\theta_4 + \theta_3}{2}$$

But $$K = \frac{Q \times 2.3026 \times \log_{10}\left(\dfrac{r_2}{r_1}\right)}{2\pi l\left(\theta_1 \dfrac{\theta_3 + \theta_4}{2}\right)} t \qquad \text{...(i)}$$

Substituting the values of Q in equation (i)

$$K = \frac{m(\theta_4 - \theta_3) \times 2.3026 \times \log_{10}\left(\dfrac{r_2}{r_1}\right)}{2\pi l\left(\theta_1 - \dfrac{\theta_3 + \theta_4}{2}\right) t} \qquad \text{...(ii)}$$

Hence, K can be determined.

HEAT HOW THROUGH A COMPOUND WALL

Consider a compound wall (or a slab) made of two materials A and B of thickness d_1 and d_2. Let K_1 and K_2 be the co-efficients of thermal conductivity of the two materials θ_1 and θ_2 are the temperatures of the end faces ($\theta_1 > \theta_2$) and 6 is the temperature of the surface in contact. After the steady state is reached, the heat flowing per second (Q) across any cross-section is the same.

For the material A, $$Q = \frac{K_1 A(\theta_1 - \theta)}{d_1}$$

For the material B,

$$Q = \frac{K_2 A(\theta - \theta_2)}{d_2}$$

From equations (i) and (ii),

$$\frac{K_1 A(\theta_1 - \theta)}{d_1} = \frac{K_2 A(\theta - \theta_2)}{d_2}$$

$$\theta = \frac{\dfrac{K_1\theta_1}{d_1} + \dfrac{K_2\theta_2}{d_2}}{\dfrac{K_1}{d_1} + \dfrac{K_2}{d_2}}$$

Substituting the value of 6 in equation (i),

$$Q = \frac{A(\theta_1 - \theta_2)}{\dfrac{d_1}{K_1} + \dfrac{d_2}{K_2}} \qquad \text{...(iii)}$$

In general, for any number of walls or slabs,

$$Q = \frac{A(\theta_1 - \theta_2)}{\Sigma\left(\dfrac{d}{K}\right)} \qquad \text{...(iv)}$$

STATISTICAL POSTULATES

Following postulates are made for applying the statistical method to physical systems:

(i) Every system has a large number of particles/molecules which are in constant random motion.

(ii) The motion of particles/molecules is described in 2Nf-dimensional phase-space.

(iii) The phase-space is divided into cells.

(iv) Postulate of equal a priori probability holds good.

(v) Probability of a particular configuration is given by the ratio of microstates corresponding to that configuration to total accessible microstates, *i.e.*,

$P_i = \Omega_i/\Sigma\Omega_i$ and the average value of any physical parameter is given by

$$< x > = \Sigma x_i P_i = \frac{\Sigma x_i \Omega_i}{\Sigma \Omega_i}$$

where Qi is the frequency of the configuration.

(vi) The equilibrium state of the system is the state of maximum probability.

(vii) The law of conservation of mass is generally valid whence

$$n_1 + n_2 + n_3 + ... = \Sigma n_i, = N = \text{constant.}$$

(viii) The law of conservation of energy governs the distribution of particles among cells. Thus, if ε_1, ε_2, ε_3 ... denotes energies of each particle in cell 1, 2, 3, etc., then

$$E = \varepsilon_1 n_1, + \varepsilon_2 n_2 + \varepsilon_3 n_3 + ... = \Sigma \varepsilon_i n_i = \text{constant.}$$

(ix) Every point in the phase-space (cell) represents a microstate, the cell volume being h where/is the degree of freedom.

(x) The number of microstates in the energy interval e and $\varepsilon + d\varepsilon$, per unit volume is represented as $\Omega(\varepsilon)$.

(xi) A collection of accessible microstates form a statistical ensemble.

Note: A complexion/configuration/arrangement represents distribution of molecules in different cells, and their form is governed by constraints stated in (vii) and (viii).

ACCESSIBLE MICROSTATES

Let us now count the number of energy states available in a macrosystem. Let E be the total energy of this system which has been divided into small energy intervals δ E so that a sufficient number of quantum stales are available in each interval of δ E. Now, if Ω (E) is the number density of such states in the energy interval E and E + δE then

$$\Omega(E) = \rho(E)\ \delta E \quad ...(i)$$

where ρ (E) is density of states (the number of states in unit energy interval). Thus

$$\Omega(E) \propto \delta E \quad ...(ii)$$

Let the value of Ω(E) changes with the magnitude of energy interval. Hence we may consider Ω(E) as a smoothly varying function of E.

Again, if ϕ(E) denotes the number of states with energy up to E and ϕ(E + δE) denotes the number of states with energy up to E + δE then

$$\Omega(E) = \phi(E + \delta E) - \phi(E) \equiv \frac{\delta\phi}{\delta E}\delta E \qquad ...(iii)$$

Comment on ρ (E). Relations (i) and (ii) (possibly) indicate that $\rho(E)$ is a constant. But it is not so. In fact ρ (E) is an increasing function of E. Let us see how! We know that the total number of microstates is equal to the product of thermodynamic probabilities of different macrostates (composite probability is obtained, always by multiplying probabilities for different complexions), *i.e.*,

$$\Omega = \Omega_1 \Omega_2 \Omega_3 \ldots \Omega_i \ldots \Omega_f$$

or

$$\Omega = \prod_1 \Omega_i \text{ (i = 1 to f)}$$

where f is the degree of freedom.

But Ω_i should be somehow related to E, *i.e.*,

$$\Omega \propto E^a \quad \therefore \Omega \propto E^{fa} \qquad ...(iv)$$

usually a $\cong$ 1/2 and/ranges between 10^{22} and 10^{28}. This suggests that for a certain energy interval δE, $\rho(E)$ increases with the value of E.

STATISTICAL MECHANICS

Statistical Mechanics deals with systems consisting of many particles and the methods employed are to get a collective or macroscopic property of the system without taking into account the individual motion of the particles The word particle is used in a broad sense meaning a fundamental particle *e.g.*, an electron, an atom, a molecule etc. The particle is a well defined and a stable unit of a given physical system. As it is not practically possible to determine the property of each particle individually, a statistical approach is made by using the concept of probability of distribution. This statistical approach helps in determining the bulk or macroscopic property of the system as a whole. The idea of probability does not imply that the particles move in a random way without obeying the laws. For a many particle system, the statistical analysis is valid regarding the applications of probability distribution of particles.

MAXWELL-BOLTZMANN DISTRIBUTION AND IDEAL GAS

Most of the gases obey Maxwell-Boltzmann distribution over a wide range of temperature. Consider a gas having mono-atomic molecules. It is assumed that the gas is ideal and has only kinetic energy of translation. Intermolecular attraction is assumed to be absent. If the kinetic energy is not quantized but considered to possess continuous values of energy, then the partition function Z can be written as

$$Z = \int_0^\infty e^{-(E/kT)}\, g(E)dE$$

Here g_i is replaced by g (E) dE

But $$g(E)dE = \left(\frac{4\pi V(2m^3)^{1/2}}{h^3}\right)E^{1/2}dE$$

Here V is the volume occupied by the gas and h is Planck's constant.

$$Z = \frac{4pV(2m^3)^{1/2}}{h^3}\int_0^\infty E^{1/2}e^{-(E/kT)}dE$$

$$Z = \frac{4\pi V(2m^3)^{1/2}}{h^3}\left[\frac{\sqrt{\pi(kT)^3}}{2}\right]$$

$$Z = \frac{V(2\pi mkT)^{3/2}}{h^3} \qquad \text{...(i)}$$

Equation (i) represents the partition function for an ideal mono-atomic gas in terms of the volume and temperature of the gas.

Taking logarithms on both sides

$$\log_e Z = \log_e\left[\frac{(2\pi m)^{3/2}\,V}{h^3}\right] + \log_e (kT)^{3/2}$$

$$\log_e Z = C + \frac{3}{2}\log_e kT \qquad \text{...(ii)}$$

Also $$E_{av} = (kT^2)\frac{d}{dT}(\log_e Z)$$

$$E_{av} = (kT^2)\frac{d}{dT}\left[C + \frac{3}{2}\log_e kT\right]$$

$$= (kT^2)\frac{d}{dT}\left[C + \frac{3}{2}\log_e k + \frac{3}{2}\log_e T\right]$$

$$E_{av} = \frac{3}{2}kT \qquad \text{...(iii)}$$

The total energy,

$$U = NE_{av}, \quad U = N\left(\frac{3}{2}kT\right)$$

$$U = \frac{3}{2}NkT. \qquad \text{...(iv)}$$

Equation (iv) shows that the internal energy of an ideal monoatomic gas depends only on its temperature. The same relation does not however, hold good for real gases. In the case of real gases the internal energy is partly potential and partly kinetic. The total energy depends on the volume of the gas.

QUANTUM STATISTICS

In Maxwell-Boltzmann distribution, it has been assumed that all the energy levels are accessible to all the particles of the system. However, there may be certain levels prohibited to a certain group of particles. Each energy state is associated with a certain available wave function. The probability of a particular distribution is restricted by the available wave functions of any state. These restrictions are taken into account in quantum statistics. There are two types of quantum statistics

(i) Fermi-Dirac statistics

(ii) Bose-Einstein statistics.

In Fermi-Dirac statistics the particles are assumed to obey Pauli's exclusion principle. They are characterised by antisymmetric wave functions. The particles under, this category are called *fermions*. Protons neutrons and electrons are fermions.

In the case of Bose-Einstein statistics, it is assumed that the particles are not restricted by Pauli's exclusion principle and are characterised by symmetric wave functions. These particles are called *bosons*. It has been experimentally found that all particles having spin zero or integral multiple of 1 are bosons. Helium nuclei and mesons are bosons. In both the kinds of quantum statistics, the particles are identical and indistinguishable. At high temperature and low pressure, all the three statistics give practically the same result.

FERMI-DIRAC DISTRIBUTION LAW

Consider a system consisting of a large number of particles. It is assumed:

(i) that the particles are *identical* and *indistinguishable*, and

(ii) that the particles obey exclusion principle. It means that no two particles can have the same dynamical state and the wave function of the whole system must be antisymmetric. The particle satisfying these conditions are called *fermions*. In general, all fundamental particles with spin are fermions.

In quantum statistics the intrinsic probability g_i is governed by the different quantum states relating to a given energy *i.e.*, the degeneracy of the energy state. Each quantum state corresponds to a particle wave function. The Wave functions are determined by each of the possible arrangements of quantum number corresponding to a given energy level. For particles with spin 1/2, in the absence of magnetic forces, each of the particle may be in the energy states with spin +1/2 or –1/2. Hence the intrinsic probability g_i in this case is 2. For motion in a central field, the energy of the particle is independent of the orientation of the orbital angular momentum Due to this, a degeneracy of $(2l + 1)$ is introduced and it is the value of g for that particular energy state. If the particles possess spin, the total degeneracy = $2(2l + 1)$. As no two particles can be in the same energy state having the same quantum number, the instrinsic probabilities g_i's give the maximum number of particles that can be accommodated in a particular energy level without violating the exclusion principle *i.e.*,

$$n_i \leq g_i$$

It means that the n_i value for a given distribution does not exceed the corresponding value of g_i.

Let n_i, be the number of particles for energy level E_i The first particle can be placed in any one of the available g_i states *i.e.*, this particle can be assigned to any of the gi sets of quantum numbers. Thus, the first particle can be distributed in g_i different ways. Similarly, the second particle can be arranged in $(g_i - 1)$ different ways and the process continues.

Thus, the total number of different ways of arranging n_i particles among the available g_i states with energy level E_i is

$$= g_i(g_i - 1)(g_i - 2) \text{......}[g_i - (n_i - 1)]$$

$$= \frac{g_i!}{(g_i - n_i)!} \quad \text{...(i)}$$

Further, if the particles are taken to be indistinguishable, it will not be possible to detect any difference when n_i particles are reshuffled into different states occupied by them in the energy level E_i. Therefore, the total number of different and distinguishable ways is,

$$= \frac{g_i!}{n_i!(g_i - n_i)!} \quad \text{...(ii)}$$

Therefore, the total number of different and distinguishable ways of getting the distribution n_1, n_2, n_3, etc., among the various energy levels, E_1, E_2, E_3...,etc.. can be obtained by multiplying the various factors.

$$\therefore \quad P = \frac{g_1!}{n_1!(g_1 - n_1)!} \cdot \frac{g_{2i}!}{n_2!(g_2 - n_2)!} \cdots$$

$$P = \prod_i \frac{g_i!}{n_i!(g_i - n_i)!} \quad \text{...(iii)}$$

The most probable distribution can by obtained by evaluating the maximum value of $\log_e$, P in equation (iii).

This should also satisfy the condition that

$$\sum_i n_i = N$$

and $\sum_i n_i \; E_i = U$.

According to Stirling's approximation

$$\log_e x! = x \log_e x - x$$

From equation (iii), applying Stirling's approximation,

$$\log_e P = \sum_i [(g_i \log_e g_i - g_1) - (n_i \log_e n_i - n_i)$$

$$- [(g_i - n_1) \log_e (g_i - n_1) - (g_i - n_1)]$$

$$\log_e P = \sum_i [g_i \log_e g_i - n_1 - \log_e n_i - (g_i - n_i)$$

$$\log_e (g_i - n_1)] \quad \text{....(iv)}$$

Differentiating equation (iv)

$$-d (\log_e P) = \sum_i [\log_e n_i - \log_e (g_i - n_i)] dn_i$$

To obtain the maximum value of P,

$$d(\log_e P) = 0$$

$$\therefore \quad \sum_i [\log_e n_i - \log_e (g_i - n_i)] dn_i = 0 \qquad ...(v)$$

But $\sum_i dn_i = 0$ ...(vi)

and $\sum_i E_i dn_i = 0$ ...(vii)

Multiplying (vi) by α and (vii) by β and adding to equation (v), we get

$$\sum_i [\log_e n_i - \log_e (g_i - n_1) + \alpha + \beta E_i] dn_i = 0$$

The equilibrium distribution is possible if

$$\therefore \quad \log_e n_i - \log_e (g_i - n_1) + \alpha + \beta E_i = 0 \qquad ...(viii)$$

$$\log_e \left(\frac{n_i}{g_i - n_i} \right) = -\alpha + \beta E_i$$

$$\frac{n_i}{g_i - n_i} = e^{-\alpha - \beta Ei}$$

$$\frac{g_i - n_i}{n_i} = e^{\alpha + \beta Ei}$$

$$\frac{g_i}{n_i} - 1 = e^{\alpha + \beta Ei}$$

$$\frac{g_i}{n_i} = (e^{\alpha + \beta Ei}) + 1$$

$$n_i = \left(\frac{g_i}{e^{\alpha + \beta E_i}} \right) + 1 \qquad ...(ix)$$

Equation (ix) represents the Fermi-Dirac distribution law. The parameter has the same role as in the case of Maxwell-Boltzmann distribution law *i.e.*, for the system consisting of fermions in statistical equilibrium.

$$\beta = \frac{1}{kT}$$

$$\therefore \quad n_i = \frac{g_i}{[e^{(\alpha + E_i/kT)}] + 1}$$

In most cases the value of a is negative and is taken to be equal to

$$\frac{-E_F}{kT}$$

$$n_i = \frac{g_i}{\left[e^{(E_i - E_F)/kT)}\right] + 1}$$

The value of E_F is positive and is independent of temperature.

For T = 0, all the energy states are fully occupied and $n_i = g_i$ All the states with $E > E_F$, are empty *i.e.*, $n_i = 0$.

$$\left[e^{(E_i - E_F)/kT)}\right] = \begin{cases} 0 \text{ for } E_i - E_F < 0 \\ \infty \text{ for } E_i - E_F > 0 \end{cases}$$

Limit $T \to 0$

In the case of Fermi-Dirac statistics, the accumulation of particles at the ground level is not allowed and at temperature T = 0, the particles occupy the lowest energy levels upto E_F. Here, the energy E_F gives the indication of the maximum energy of the fermions in the system. By is also called Fermi energy. For higher temperatures, the particles, occupy higher energy states greater than E_F. The curves indicate that only those fermions with energies close to E_F can move into unoccupied higher energy states. If $k\theta_F = E_F$, the temperature θ_F is called the Fermi temperature.

STATISTICAL EQUILIBRIUM

Consider that a thermodynamical isolated system consists of N particles. The energy states available to the particles are E_1, E_2, E_3 etc. These energy states may be quantized or may be continuous and are due to vibrational and rotational energy of the particles.

Suppose, that at any given instant of time, n_1 particles are in state of energy E_1, n_2, particles with energy state E_2 and so on.

The total number of particles in the system is,

$$N = n_1 + n_2 + n_3 + \ldots\ldots = \sum_i n_i \qquad \ldots(i)$$

Here $\qquad i = 1, 2, 3$ etc.

The total energy of the system,

$$U = n_1E_1 + n_2E_2 + n_3E_3 + ...$$

$$U = \sum_i n_i E_i \qquad ...(ii)$$

Equation (ii) refers to the total energy of a system in which the particles are non-interacting. Here, the energy of each particle depends only on the coordinates of the particle in the system. For an isolated system, the total energy U is constant.

For isolated system.

$$U = \sum_i n_i E_i = \text{constant} \qquad ...(iii)$$

Consider a gas having N molecules at a certain temperature and pressure. Its volume, temperature and pressure are kept constant *i.e.*, the system is isolated. The total energy of this system remains constant.

But, the molecules of the gas collide with each other and also with the walls of the container. Consequently, the number of molecules change from one energy state to the other energy state. It means that the value of n_1, n_2, n_3 etc., continuously change. It can be reasonably assumed that for each microscopic State of a system of particles, there is a particular most favoured distribution.

When this distribution or partition is reached, the system attains statistical equilibrium. For an isolated system, the values of n_1, n_2, n_3 etc., vary only near the values corresponding to the most probable distribution.

PROBABILITY THEOREMS IN STATISTICAL THERMODYNAMICS

The following are the important probability theorems commonly used in statistical thermodynamics:

(1) The number of ways in which N distinguishable particles can he arranged in order is equal to

$$N!$$

(2) The number of different ways in which N particles can be selected from N distinguishable particles irrespective of the order of selection is equal to

$$\frac{N!}{(N-n)!n!}$$

(3) The number of different ways in which n indistinguishable particles can be arranged in g distinguishable states with not more then one particle in each state is equal to

$$\frac{g!}{(g-n)!n!}$$

MAXWELL-BOLTZMANN DISTRIBUTION LAW

Consider a system that contains a large number of particles that are identical and distinguishable. The identical particles refer to the particles having the same structure. The distinction between the particles is due to their energy states at a given instant. Four particles are in energy state E_1, 2 in state E_2, zero in state E_3 and so on. It is assumed that all the energy states are accessible to each particle. Consequently, it can be assumed that the probability of any particular partition is proportional to the number of different ways in which the particles can be distributed in the existing available energy states so as to produce the desired partition. The first particle a in state E_1 can be selected in N ways. The second particle b in state E_1 can be selected in (N – 1) ways and so on. Therefore, the total number of ways in which the first four particles in state E_1 can be selected is given by,

$$N(N-1)(N-2)(N-3) = \frac{N!}{(N-4)!}$$

Moreover, the four particles in state E_1 can be arranged in 4! different orders. For example, *abcd, bcda, cdab* and so on. There are 24 ways. But, it is immaterial for these particles to be arranged in any particular order in state E_1, because they are identical. Thus, the total number of *distinguishable* different ways are,

$$\frac{N!}{4!(N-4)!}$$

In general, if the first state consists of n_1 particles, the distinguishable different ways, for arranging n_1 particles in state E_1, are,

$$p_1 = \frac{N!}{n_1!(N-n_1)!} \quad \text{...(i)}$$

For the second state E_2, only $(N - n_1)$ particles are available and n_2 particles are in state E_2. The number of distinguishable different ways are,

$$p_2 = \frac{(N-n_1)!}{n_2!(N-n_1-n_2)!} \qquad ...(ii)$$

If this process is continued for all the available states, the total number of distinguishable ways are obtained by multiplying P_1, P_2, P_3 etc.

$$P = p_1 \times p_2 \times p_3 \times \ ......$$

$$P = \left[\frac{N!}{n_1!(N-n_1)!}\right]\left[\frac{(N-n_1)!}{n_2!(N-n_1-n_2)}\right] \times ....$$

$$P = \frac{N!}{n_1!n_2!n_3!...} \qquad ...(iii)$$

The distinguishable ways are,

$$P = \frac{N!}{4!2!0!3!1!} \qquad ...(iv)$$

Here 0! is equal to one.

It has been assumed so far that all the available states have the same probability of occupation by the particles. However, it may happen that the states have different intrinsic probabilities say g_i. For example, a particular energy state may be favourable with more different angular momentum states than the rest and hence it is more likely to be occupied. Taking into account this intrinsic probability factor, the value of P will be different.

If g_i is the probability of locating a particle in a certain energy state E_i, then the probability of locating 2 particles in the same state is $g_i \times g_i = g^2_i$. For n_i particles, the probability is $g_i n_i$. Hence the total probability for a given distribution is given by

$$P = \frac{N! g_1 n_1 g_2 n_2 g_3 n_3 ...}{n_1!n_2!n_3!n_4!...} \qquad ...(v)$$

Here n_1, n_2, n_3 etc., are the number of particles in states E_1, E_2, E_3, etc. and g_1, g_2, g_3 etc., are the intrinsic probabilities for states E_1, E_2, E_3 etc.

If all the particles are further assumed to be indistinguishable *i.e.*, particles in state E, and particles in state E_4 cannot be distinguished, then N! permutations among the particles themselves and occupying the different states result in the same distribution. The probability in this case is given by

$$P = \frac{1}{N!}\left[\frac{N! g_1 n_1 g_2 n_2 g_3 n_3 \ldots}{n_1! n_2! n_3! \ldots}\right]$$

$$\therefore \quad P = \frac{g_1 n_1 g_2 n_2 g_3 n_3 \ldots}{n_1! n_2! n_3! \ldots}$$

$$\therefore \quad P = \prod_i^N \frac{g_i n_i}{n_1!} \qquad \ldots(vi)$$

Here Π is the p: oduct sign (since probability is the product of such distributions.

The most probable distribution can be obtained by evaluating the maximum value of $\log_e$, P in equation (vi). This should also satisfy the two conditions that

$$\sum_i n_i = N \qquad \ldots(vii)$$

and $\sum_i n_i \; E_i = U$ ...(vii)

According to Stirling's approximation

$$\log_e x! = x \log_e x - x$$

From equation (vi), applying Stirling's approximation, we get

$$\log_e P = \sum_i (n_i \log_e g_i - \log_e n_i!)$$

$$= \sum_i [n_i \log_e g_i - (n_i \log_e n_i - n_i)]$$

$$= \sum_i (n_i \log_e g_i - n_i \log_e n_i + n_i)$$

$$= \sum_i n_i - \sum_i (n_i \log_e n_i - n_i \log_e g_i)$$

$$\log_e P = N \sum_i n_i \log_e \frac{n_i}{g_i} \qquad \ldots(ix)$$

Differentiating equation (ix)

$$d(\log_e P) = \sum_i (dn_i) \log_e \frac{n_i}{g_i} - \sum_i n_i d\left(\log_e \frac{n_i}{g_i}\right)$$

$$= \sum_i dn_i \log_e \frac{n_i}{g_i} - \sum_i n_i \frac{dn_i}{n_i}$$

$$= \sum_i dn_i \log_e \frac{n_i}{g_i} - \sum_i dn_i$$

But $\sum_i dn_i = 0$.

$$\therefore \quad -d(\log_e P) = \sum_i \left[\log_e \left(\frac{n_i}{g_i} \right) \right] dn_i \qquad ...(x)$$

To obtain the maximum value of P,

$$d(\log_e P) = 0$$

$$\therefore \quad \sum_i \log_e \left(\frac{n_i}{g_i} \right) dn_i = 0 \qquad ...(xi)$$

But $\Sigma dn_i = 0$ ...(xii)

and $\Sigma E_i dn_i = 0$ ...(xiii)

Multiplying (xii) by α and (xiii) by β and adding to equation (xi), we get

$$\sum_i \left[\log_e \left(\frac{n_i}{g_i} \right) + \alpha + \beta E_i \right] dn_i = 0$$

The equilibrium distribution is possible if

$$\log_e \frac{n_i}{g_i} + \alpha + \beta\, Ei = 0 \qquad ...(xiv)$$

or $$\frac{n_i}{g_i} = e^{-\alpha-\beta Ei}$$

or $$n_i = g_i\, e^{-\alpha-\beta Ei} \qquad ...(xv)$$

This gives the maximum probability distribution and a and p are two parameters that depend upon the physical property of the system.

$$N = n_1 + n_2 + n_3 + ...$$

$$N = g_1 e^{-\alpha-\beta E1} + g_1 e^{-\alpha-\beta E2} + ...$$

$$N = e^{-\alpha}[g_1 e^{-\beta E1} + g_2 e^{-\beta E2} + ...]$$

$$N = e^{-\alpha}[\sum_i g_i\, e^{-\beta Ei}]$$

Take $[\sum_i g_i\, e^{-\beta Ei} = Z$ ...(xvi)

Here, Z is called the partition function.

$$N = e^{-\alpha}(Z) \qquad \text{....(xvii)}$$

$$e^{-\alpha} = \frac{N}{Z}$$

Substituting this value in equation (vii)

$$n_i = \left(\frac{N}{Z}\right)[g_i e^{-\beta E_i}] \qquad \text{....(xviii)}$$

This equation refers to Maxwell-Boltzmann Distribution Law.

MAXWELL-BOLTZMANN DISTRIBUTION IN TERMS OF TEMPERATURE

The total energy of an isolated system is given by

$$U = \sum_i n_i E_i$$

$$U = n_1E_1 + n_2E_2 + n_3E_3 + ...$$

$$U = (g_1e^{-\alpha-\beta E1}) E_1 + (g_2e^{-\alpha-\beta E2}) E_2 + ...$$

$$U = e^{-\alpha}[g_1E_1e^{-\beta E1} + g_2E_2e^{-\beta E2} + ...]$$

But $\quad e^{-\alpha} = \frac{N}{Z}$

$$\therefore \qquad U = \frac{N}{Z}[g_1E_1e^{-\beta E1} + g_2E_2e^{-\beta E2} + ...]$$

$$U = \frac{N}{Z}\sum_i g_i E_i e^{-\beta Ei} \qquad \text{...(i)}$$

Here $\quad Z = \sum_i g_i e^{-\beta Ei} \qquad \text{...(i)}$

$$\frac{dZ}{d\beta} = \frac{d}{d\beta}\sum_i g_i e^{-\beta Ei} = -\sum_i g_i E_i e^{-\beta Ei}$$

$$\therefore \qquad \sum_i g_i E_i e^{-\beta Ei} = \frac{d}{d\beta}\sum_i g_i E_i e^{-\beta Ei}$$

Substituting this value in equation (i)

$$U = -\frac{N}{Z}\frac{d}{d\beta}\sum_i g_i e^{-\beta Ei}$$

$$U = -\frac{N}{Z}\frac{d}{d\beta}$$

$$U = - N\frac{d}{d\beta}[\log_e Z] \qquad ...(ii)$$

The average energy of a particle is

$$E_{av} = -\frac{U}{N} = -\frac{d}{d\beta}(\log_e Z) \qquad ...(iii)$$

This shows that for a given system, the total energy U, the partition function Z, and the average energy of the particle E_{av}, depend on the parameter β. Therefore, β may be taken to characterise the internal energy of the system and has the units per joule. If T is the temperature in degrees kelvin, then it is more convenient to represent parameter β as

$$\beta = \frac{1}{kT} \text{ or } KT = \frac{1}{\beta} \qquad ...(iv)$$

Here kT has the unit of energy *i.e.*, joule and k is the Boltzmann's constant. Its units are J/K. The value of k is given by

$$k = 1.3805 \times 10^{-23} \text{ J/K}$$

Substituting the value of

$$\beta = \frac{1}{kT}$$

in all the equations, we get

$$Z = \sum_i g_i e^{-(Ei/kT)} \qquad ...(v)$$

$$n_i = \frac{N}{Z}\sum_i g_i e^{-(Ei/kT)} \qquad ...(vi)$$

This equation represents the Maxwell-Boltzmann distribution law in terms of the temperature of the system.

As $\beta = \frac{1}{kT}$

$$d\beta = -\frac{dT}{kT^2}$$

Substituting this value in equation (ii)

$$\therefore \quad U = (kNT^2)\frac{d}{dT}(\log_e Z) \qquad ...(vii)$$

Also, from equation (iii)

$$E_{av} = (kT^2)\frac{d}{dT}(\log_e Z) \qquad ...(viii)$$

Equation (viii) gives the relation between the average energy of the particle and its temperature under equilibrium position. Hence, the temperature of a system in statistical equilibrium is the physical quantity related to the average energy of the particle of the system.

SOME SOLVED PROBLEMS

Problem 1:

For N = 400, find the ratios of the probabilities for (300, 100) and (200, 200) distribution in two identical boxes.

Solution:

We have

$$P(300, 100) = \frac{400!}{300!\,100!}\,\frac{1}{2^{400}}$$

$$P(200, 200) = \frac{400!}{200!\,200!}\,\frac{1}{2^{400}}$$

$$\frac{P(300, 100)}{P(200, 200)} = \frac{200!\,200}{300!\,100!} \cong 10^{-5}$$

Problem 2:

Eight similar coins are tossed, find the probability of three heads and t tails.

Solution:

We know

$$P(3, 5) = \frac{^8C_3}{2^8} = \frac{8\times7\times6}{3\times2\times1\times2^8}$$

or $$P(3, 5) = \frac{7}{32}$$

Problem 3:

A system consists of 6000 particles distributed in three energy states with equal spacing. The energy of the three states are $E_{l'} = 0$, $E_{2'} = x$ and $E_3 = 2x$. All the three slates have the same intrinsic probability g.

At a certain instant there are 3000 particles in lower level, 2500 in the middle level and 500 in the upper level. Compare the relative probabilities with the distribution obtained, by the transfer of one particle from the middle to the lower level and one particle from the middle to the upper level and the original distribution.

Solution:

Let P_1 and P_2 be the probabilities in the two cases

In the first case,

$$N = 6000,$$

$$n_1 = 3000,$$

$$n_2 = 2500,$$

$$n_3 = 500.$$

$$P_1 = \frac{g^N}{n_1!n_2!n_3!}$$

$$P_1 = \frac{g^{6000}}{3000!2500!500!} \quad ...(i)$$

In the second case

$$N = 6000,$$

$$n_1 = 3001,$$

$$n_2 = 2498,$$

$$n_3 = 501.$$

$$P_1 = \frac{g^N}{n_1!n_2!n_3!}$$

$$P_1 = \frac{g^{6000}}{3001!2498!501!}$$

Dividing (ii) by (i)

$$\frac{P_2}{P_1} = \frac{3000!2500!500!}{3001!2498!501!}$$

$$\frac{P_2}{P_1} = \frac{2500 \times 2499!}{3001 \times 501!}$$

$$\frac{P_2}{P_1} = 4.157$$

$$\approx 4.2.$$

It means the transfer of one particle from the middle to the upper and one particle from the middle to the lower state has changed the probability by a factor 4.2. This shows that both these distributions are not near the equilibrium state.

Problem 4:

In an Ingen-hausz experiment, wax melted over 10 cm of copper rod and over 4 cm of iron rod. What is the conductivity of iron when the conductivity of copper is 0.90?

Solution:

Here,

$$l_1 = 10\text{cm}$$

$$l_2 = 4\text{cm}$$

$$K_1 = 0.90$$

$$K_2 = ?$$

$$\frac{K_2}{K_1} = \frac{l_2^2}{l_1^2}$$

$$K_2 = \frac{l_2^2}{l_1^2} \times K_1$$

$$= \frac{16}{100} \times 0.90$$

$$\mathbf{K_2 = 0.144.}$$

Problem 5:

The opposite faces of a metal plate of 0.2 cm thickness an at a difference of temperature of 109°C and the area of the plate is 200 sq cm. Find the quantity of heat that will flow through the plate in one minute if K = 0.2 CGS units.

Solution:

Here,

$$K = 0.2$$

$$A = 200 \text{ sq cm}$$

$$d = 0.2 \text{ cm}$$

$$(\theta_1 - \theta_2) = 100°C$$

$$t = 60s$$

$$Q = \frac{KA(\theta_1 - \theta_2)t}{d}$$

$$= \frac{0.2 \times 200 \times 100 \times 60}{0.2}$$

$$= \mathbf{12 \times 10^5 \text{ cal.}}$$

Problem 6:

A bar of length 30 cm and uniform area of cross-section 5 cm² consists of two halves AB of copper and BC of iron welded together at B. The end A is maintained at 200°C and the end C at 0°C. The sides of the bar are thermally insulated. Find the rate of flow of heat along the bar when the steady state is reached. Thermal conductivity of copper if 0.9 and thermal conductivity of iron is 0.12 CGS units.

Solution:

Suppose the temperature at the interface. B is θ after the steady state is reached.

$$\therefore \quad \frac{K_1A(200-\theta)}{d} = \frac{K_2A(\theta-0)}{d}$$

$$0.9(200 - \theta) = 0.12\,\theta$$

$$\theta = 176.5°C.$$

After the steady state is reached, the rate of flow of heat is the same in both the bars.

$$\therefore \quad Q = \frac{K_1A(\theta-0)}{d}$$

$$Q = \frac{0.9 \times 5 \times (200-176.5)}{15}$$

$$\mathbf{Q = 7.05 \text{ cal/s.}}$$

Problem 7:

An ice box is built of wood 1.76 cm thick, lined inside with cork 3 cm thick. If the temperature of the inner surface of the cork is 0°C and that of the outer surface of wood is 12°C, what is the temperature of the interface ? The thermal conductivity of wood and cork are 0.0006 and 0.00012 CGS units respectively.

Solution:

Suppose the temperature of the interface is 9 after the steady state is reached

$$\therefore \quad \frac{K_1A(12-\theta)}{d_1} = \frac{K_2A(\theta-0)}{d_2}$$

Here
$$K_1 = 0.0006,$$
$$K_2 = 0.00012$$
$$d_1 = 1.75 \text{ cm},$$
$$d_2 = 3 \text{ cm}$$

$$\frac{0.0006(12-\theta)}{1.75} = \frac{0.00012(\theta)}{3}$$

$$\theta = \mathbf{10.74°C.}$$

Problem 8:

Find the time in which a layer of ice 3 cm thick on the surface of a pond will increase its thickness by 1 mm when the temperature of the surrounding air is –20°C.

Thermal conductivity of ice = 0.005

Latent heat of ice = 80 cal/g

Density of ice at 0°C = 0.91 g/cm³.

Solution:

$$\int dt = \frac{\rho L}{K\theta}\int x dx$$

$$t = \frac{\rho L}{2K\theta}[x_2^2 - x_1^2]$$

Here,
$$\rho = 0.91 \text{ g/cm}^3$$
$$L = 80 \text{ cal/g}$$
$$K = 0.005$$
$$\theta = 20°C$$
$$x_1 = 3 \text{ cm}$$

$$x_2 = 3.1 \text{ cm}$$

$$t = \frac{0.91 \times 80}{2 \times 0.005 \times 20} [(3.1)^2 - (3)^2]$$

$$= 222.04 \text{ s}$$

$$= \textbf{3 min 42 s approximately.}$$

Problem 9:

How much time will it take for a layer of ice of thickness 10 cm to increase by 5 cm on the surface of a pond when the temperature of the surroundings is –10°C?

Solution:

$$K = 0.005$$

$$L = 80 \text{ cal/g}$$

$$p = 0.90 \text{ g/cm}^3$$

Here $$\int dt = \frac{\rho L}{K\theta} \int x dx$$

$$t = \frac{\rho L}{2K\theta} [x_2^2 - x_1^2]$$

Here $$x_1 = 10 \text{ cm},$$

$$x_2 = 15 \text{ cm}$$

$$t = \frac{0.90 \times 80}{2 \times 0.005 \times 10} [(15)^2 - (10)^2]$$

$$= \mathbf{9 \times 10^4 s = 25 \text{ hours.}}$$

Problem 10:

Two large closely spaced concentric spheres (both are black body radiators) are maintained, at temperatures of 200 K and 300 K respectively. The space in between the two spheres is evacuated. Calculate the net rate of energy transfer between the two spheres. (σ = 5.672 × 10^{-8} M.K.S. units).

Solution:

Here $$T_1 = 300 \text{ K}$$

$$T_2 = 200 \text{ K}$$

$$\sigma = 5.672 \times 10^{-8} \text{ M.K.S. units}$$

$$R = \sigma(T_1^4 - T_2^4)$$

$$= 5.672 \times 10^{-8}[(300)^4 - (200)^4]$$

$$\mathbf{R = 368.69 \text{ watts/m}^2.}$$

Problem 11:

Calculate the radiant emittance of a black body at a temperature of (i) 400 K (ii) 4000 K. (σ = 5.672 × 10^{8} M.K.S. units).

Solution:

Here $T = \sigma T^4$

(i) $R = 5.672 \times 10^{-8} [400]^4$

$= \mathbf{1452 \text{ watt/m}^2.}$

(ii) $R = 5.672 \times 10^{-8}(4000)^4 = 1452 \times 10^4 \text{ watts/m}^2$

$= \mathbf{14520 \text{ kilo-watts/m}^2.}$

Problem 12:

The relative emittance of tungsten is approximately 0.35. A tungsten sphere of surface area 10^{8} sq metres is suspended inside a large evacuated enclosure whose walls are at 300 K. What power input is required to maintain the sphere at a temperature of 3000 K? The conduction of heat along the supports can be neglected. (σ = 5.672 × 10^{8} M.K.S. units).

Solution:

Here $R = \sigma A e[T^4 - T_0^4]$

$A = 10^{-3}$ sq meters

$e = 0.35$

$\sigma = 5.672 \times 10^{-8}$ M.K.S. units

$R = 5.672 \times 10^{-8} \times 10^{-3} \times 0.35[(3000)] - (300)^4]$

$= \mathbf{1609 \text{ watts}}$

The power input = **1609 watts.**

Problem 13:

An aluminium foil of relative emittance 0.1 is placed in between two concentric spheres at temperatures 300 K and 200 K respectively. Calculate the temperature of the foil after the steady state is reached. Assume that the spheres are perfect black body radiators. Also calculate the rate of energy transfer between one of the spheres and the foil. ($\sigma = 5.672 \times 10^{-8}$ M.K.S. units).

Solution:

Here
$$T_1 = 300 \text{ K}$$
$$T_2 = 200 \text{ K}$$
$$e = 0.1$$
$$\sigma = 5.672 \times 10^{-8} \text{ M.K.S. units.}$$

(i) Let x be the temperature of the foil. After the steady-state is reached,

$$e\sigma(T_1^4 - x^4) = e\sigma(x^4 - T_2^4)$$

or
$$[(300)^4 - x^4] = [x - 200)^4]$$
$$x^4 = 48.5 \times 10^8$$
$$\mathbf{x = 263.8\ K}$$

(ii)
$$R = e\sigma(T_1^4 - x^4)$$
$$R = 0.1 \times 5.672 \times 10^{-8}[(300)^4 - (263.8)^4]$$
$$\mathbf{R = 18.5\ watt/m^2}$$

Problem 14:

Calculate the energy radiated per minute from the filament of an incandescent lamp at 2000 K, if the surface area is 5.0×10^{-5} sq metres and its relative emittance is 0.85.

Solution:

$$E = Ae\ \sigma t\ (T^4)$$
$$A = 5 \times 10^{-5} m^2$$
$$e = 0.85$$

$\sigma = 5.672 \times 10^{-8}$ M.K.S. units.

$t = 60s$

$T = 2000$ K

$E = 5 \times 10^{-6} \times 0.85 \times 5.672 \times 10^{-8} \times 60 \times (2000)^4$

E = 2315 joules.

Problem 15:

An iron furnace radiates 1.53×10^5 calories per hour through an opening of cross-section 10^{-4} sq metre. If the relative emittance of the furnace is 0.80, calculate the temperature of the furnace. (Given $\sigma = 1.36 \times 10^{-8}$ $cal/m^2\text{-}s\text{-}K^4$).

Solution:

$E = Ae\,\sigma\, f(T^4)$

$E = 1.53 \times 10^5$ calories

$A = 10^{-4}\ m^2$

$e = 0.80$

$t = 3600$ s

$T = ?$

$$T^4 = \frac{E}{Ae\sigma t}$$

$$T = \left(\frac{E}{Ae\sigma t}\right)^{1/4}$$

$$= \left(\frac{1.53\times10^5}{10^{-4}\times0.80\times1.36\times10^{-8}\times3600}\right)^{1/4}$$

T = 2500 K.

Problem 16:

Calculate the black body temperature of the sun from the following data.

Solution:

Stefan's constant $= 1.37 \times 10^{-12}$ $cal/cm^2/s$

Solar constant = 2.3 cal/cm^2/minute

Radius of the sun = 7 × 10^{10} cm.

Distance between the sun and the earth

= 1.5 × 10^{13}cm

$$E = \frac{4\pi R^2 S}{4\pi r^2} = \left(\frac{R}{r}\right)^2 .S$$

Here R = 1.5 × 10^{13}cm

r = 7 × 10^{10}cm

S = 2.3 cal/cm^2/minute

$$E = \left[\frac{1.5\times10^{13}}{7\times10^{10}}\right]^2 \times\frac{2.3}{60}\text{cal/s} \quad ...(i)$$

$$1.37 \times 10^{-12} \times T^4 = \left[\frac{1.5\times10^{13}}{7\times10^{10}}\right]^2 \times\frac{2.3}{60}$$

$$E = \sigma T^4 \quad ...(ii)$$

Equating (i) and (ii)

T = 5987K.

5

DERIVATION OF STEFAN'S AND NEWTON'S LAWS

INTRODUCTION

The fact that blade body radiations exert pressure similar to a gas, helps in applying thermodynamics to heat radiations.

Let ψ be the energy density of radiations inside a uniform temperature enclosure at temperature T. P is the pressure and V is the volume.

Applying the first law of thermodynamics

$$\delta H = dU + P.dV \qquad \text{...(i)}$$

Applying thermodynamical relation

$$\left(\frac{\partial H}{\partial V}\right)_r = \left(\frac{\partial P}{\partial T}\right)_v \qquad \text{...(ii)}$$

$$\left(\frac{\partial U + P\partial V}{\partial V}\right)_r = T\left(\frac{\partial P}{\partial T}\right)_v$$

$$\left(\frac{\partial U}{\partial V}\right)_r = T\left(\frac{\partial P}{\partial T}\right)_v = P \qquad \text{...(iii)}$$

Now $\qquad U = V\psi$

and $\qquad P = \dfrac{\psi}{3}$

or $\qquad \left[\dfrac{\partial U}{\partial V}\right]_r = \psi$

Here ψ is a function of temperature alone.

Substituting these values in equation (iii),

$$\psi = \frac{T}{3}\frac{d\psi}{dT} - \frac{\psi}{3}$$

$$\frac{4\psi}{3} = \frac{T}{3}\frac{d\psi}{dT}$$

or $$\frac{d\psi}{\psi} = 4\frac{dT}{T}$$

Integrating, $\log \psi = 4 \log T + \text{constant}$

or $$\psi = KT^4 \qquad ...(iv)$$

Here K is a constant

Also the total rate of emission per unit area of a black body is proportional to the energy density.

$\therefore$ $$R \propto \psi \propto T^4$$

$\therefore$ $$R = \sigma T^4 \qquad ...(v)$$

where σ is Stefan's constant.

The value of Stefan's constant in C.G.S. system is 5.672×10^{-5} C.G.S. units and in M.K.S. system it is 5.672×10^{-8} M.K.S. units.

DERIVATION OF NEWTON'S LAW

Stefan's law is applicable for all temperatures of a hot body. But Newton's Law is applicable when the difference of temperature between the hot body and the surrounding is small. Consider a hot body at a temperature T_1 placed in a uniform temperature enclosure at T_1. According to Stefan's law,

$$R = e\sigma(T_1^4 - T_1^4) \qquad ...(i)$$

Here e is the emissitivity of the surface of the hot body

$$R = e\sigma(T_1 - T_2)(T_1^3 + T_1^2 T_2 + T_1 T_2^2 + T_2^3)$$

As $(T_1 - T_2)$ is small, T_1 can be taken approximately equal to T_2.

Then, $$R = e\sigma(T_1 - T_2)(T_2^3 + T_2^3 + T_2^3 + T_2^3)$$

$$R = 4e\sigma T_2^3((T_1 - T_2)$$

Taking $4e\sigma T_2^3 = k$

$$R = k(T_1 - T_2)$$

or $$R \propto (T_1 - T_2). \qquad ...(ii)$$

This equation represents Newton's law of cooling and is true when the difference of temperature is small.

Photon Gas

The interaction of electromagnetic radiations with matter, led to the idea that electromagnetic radiations are composed of discrete energy particles called photons. Each photon has an energy $h\nu$ and momentum h/λ. Here ν is the frequency and λ is the wavelength of the radiations. The electromagnetic radiations trapped in a cavity and in thermal equilibrium with the walls of the cavity are termed as black body radiations. In the equilibrium condition, the black body radiations can be considered as the *photon gas.* It is assumed that the photons do not interact among themselves. The photons interact only with-the atoms of the walls of the cavity. It is further assumed that the photons are indistinguishable and many photons can have the same energy. Photons are taken as bosons and they obey Bose-Einstein statistics.

As the photons can either be emitted or absorbed by the atoms of the walls of the cavity, the number of photons is not constant *i.e.*, the condition $\sum_i dn_i = 0$ is no longer valid.

Due to this reason the value of α is equal to zero in the Bose-Einstein distribution law.

$$n_i = \frac{g_i}{[e^{(\alpha+E_i/kT)}]-1} \qquad ...(i)$$

Here $\alpha = 0$

$$\therefore \quad n_i = \frac{g_i}{[e^{E_i/kT}]-1} \qquad ...(ii)$$

In case, the cavity is large as compared to the wavelength of the radiations, the energy spectrum of the photons is taken to be continuous. In this case, the energy difference between successive allowed energy value is very small. Thus, replacing g_i by $g(E)\,dE$, and n_i by dn

$$dn = \frac{g(E)dE}{\left(e^{E/kT)}\right)-1} \qquad ...(iii)$$

As the energy of a photon, $E = h\nu$, the value of $g(E)\,dE$ can be taken equal to $g(\nu)\,d\nu$. The factor $g(\nu)\,d\nu$ corresponds to the number of oscillatory modes in the frequency range $d\nu$ and relating to the energy range dE.

The number of states in a black body radiation in the frequency range ν and $\nu + d\nu$ can be obtained by calculating the spherical volume bound by the spheres of radii $\frac{h(\nu + d\nu)}{c}$ and $\frac{h\nu}{c}$

The volume of the spherical shell

$$= \frac{4}{3}\pi \frac{h^3}{c^3}(\nu + d\nu)^3 - \frac{4}{3}\pi \frac{h^3}{c^3}\nu^3$$

$$= \frac{4}{3}\pi \frac{h^3}{c^3}[\nu^3 + 3\nu^2 d\nu + ... - \nu^3]$$

$$= \frac{4}{3}\pi \frac{h^3}{c^3} \times 3\nu^2 d\nu$$

$$= 4\pi \frac{h^3}{c^3} \nu^2 d\nu \quad \text{...(iv)}$$

The phase space has volume $V = h^3$ and there are two states of polarization for the radiation. The number of states in the black body radiation in the frequency range ν and $\nu + d\nu$ is given by

$$g(\nu)d\nu = \frac{8\pi\nu}{c^3}\nu^2 d\nu \quad \text{...(v)}$$

$$dn = \left(\frac{8\pi V\nu^2 d\nu}{c^3}\right)\frac{1}{\left(e^{h\nu/kT}\right) - 1} \quad \text{...(vi)}$$

For dn photons in the frequency range $\nu + d\nu$ and ν, the energy is equal to $(h\nu)$ dn and the energy per unit volume

$$= \frac{(h\nu)dn}{\nu}$$

Therefore, the energy density distribution for black body radiation is given by

$$E(\nu) = \frac{(h\nu)dn}{V.d\nu} \quad \text{...(vii)}$$

Substituting the value of dn from equation (vi), we get

$$E(\nu) = \frac{8\pi h\nu^2 d\nu.h\nu}{c^3 V.d\nu\left[\left(e^{h\nu/kT}\right) - 1\right]}$$

$$E(\nu) = \left(\frac{8\pi h\nu^3}{c^3}\right)\left[\frac{1}{\left(e^{h\nu/kT}\right)-1}\right]^{d\nu} \qquad \text{...(viii)}$$

Equation (viii) represents Planck's radiation law for black body radiations.

From equation (viii)

$$E(\nu)d\nu = \frac{8\pi h\nu^3}{c^3}\left[\frac{1}{\left(e^{h\nu/kE}\right)-1}\right]^{d\nu}$$

But $\qquad \nu = \dfrac{c}{\lambda}$

or $\qquad d\nu = -\dfrac{c}{\lambda^2}d\lambda$

Neglecting the negative sign, we get

$$\therefore \qquad E(\lambda)d\lambda = \frac{8\pi h\nu^3}{c^3} \times \frac{c}{\lambda^2}\left[\frac{1}{\left(e^{h\nu/kT}\right)-1}\right]^{d\lambda}$$

or
$$E(\lambda) = \frac{8\pi hc}{\lambda^5} \times \frac{c}{\lambda^2}\left[\frac{1}{\left(e^{h\nu/kT}\right)-1}\right] \qquad \text{...(ix)}$$

This gives the energy density for wavelength λ in the spectrum of the black body. Both the equations (viii) and (ix) represent Planck's radiation law for black body radiation.

FERMI-DIRAC DISTRIBUTION LAW

Consider a system consisting of a large number of particles. It is assumed:

(i) The particles are *identical* and *indistinguishable*, and

(ii) The particles obey exclusion principle. It means that no two particles can have the same dynamical state and the wave function of the whole system must be antisymmetric. The particle satisfying these conditions are called *fermions*. In general, all fundamental particles with spin are fermions.

In quantum statistics the intrinsic probability g_i is governed by the different quantum states relating to a given energy *i.e.*, the degeneracy of the energy state. Each quantum state corresponds to a particle wave function. The Wave functions are determined by each of the possible arrangements of quantum number corresponding to a given energy level. For particles with spin 1/2, in the absence of magnetic forces, each of the particle may be in the energy states with spin +1/2 or −1/2. Hence the intrinsic probability g_i in this case is 2. For motion in a central field, the energy of the particle is independent of the orientation of the orbital angular momentum Due to this, a degeneracy of $(2\,l + 1)$ is introduced and it is the value of g for that particular energy state. If the particles possess spin, the total degeneracy = $2\,(2l + 1)$. As no two particles can be in the same energy state having the same quantum number, the instrinsic probabilities g_i's give the maximum number of particles that can be accommodated in a particular energy level without violating the exclusion principle *i.e.*,

$$n_i \leq g_i$$

It means that the n_i value for a given distribution does not exceed the corresponding value of g_i.

Let n_i, be the number of particles for energy level E_i The first particle can be placed in any one of the available g_i states *i.e.*, this particle can be assigned to any of the gi sets of quantum numbers. Thus, the first particle can be distributed in g_i different ways. Similarly, the second particle can be arranged in $(g_i - 1)$ different ways and the process continues.

Thus, the total number of different ways of arranging n_i particles among the available g_i states with energy level E_i is

$$= g_i(g_i - 1)(g_i - 2) \ldots\ldots[g_i - (n_i - 1)]$$

$$= \frac{g_i!}{(g_i - n_i)!} \qquad \ldots(i)$$

Further, if the particles are taken to be indistinguishable, it will not be possible to detect any difference when n_i particles are reshuffled into different states occupied by them in the energy level E_i. Therefore, the total number of different and distinguishable ways is,

$$= \frac{g_i!}{n_i!(g_i - n_i)!} \qquad \ldots(ii)$$

Therefore, the total number of different and distinguishable ways of getting the distribution n_1, n_2, n_3, etc., among the various energy levels, E_1, E_2, E_3...,etc.. can be obtained by multiplying the various factors.

$$\therefore \qquad P = \frac{g_1!}{n_1!(g_1 - n_1)!} \cdot \frac{g_{2i}!}{n_2!(g_2 - n_2)!} \ldots$$

$$P = \prod_i \frac{g_i!}{n_i!(g_i - n_i)!} \qquad \ldots(iii)$$

The most probable distribution can by obtained by evaluating the maximum value of $\log_e$, P in equation (iii).

This should also satisfy the condition that

$$\sum_i n_i = N$$

and $\sum_i n_i E_i = U$.

According to Stirling's approximation

$$\log_e x! = x \log_e x - x$$

From equation (iii), applying Stirling's approximation,

$$\log_e P = \sum_i [(g_i \log_e g_i - g_1) - (n_i \log_e n_i - n_i) - [(g_i - n_1) \log_e (g_i - n_1) - (g_i - n_1)]$$

$$\log_e P = \sum_i [g_i \log_e g_i - n_1 - \log_e n_i - (g_i - n_i) \log_e (g_i - n_1)] \qquad \ldots(iv)$$

Differentiating equation (iv)

$$-d (\log_e P) = \sum_i [\log_e n_i - \log_e (g_i - n_i)]dn_i$$

To obtain the maximum value of P,

$$d(\log_e P) = 0$$

$$\therefore \qquad \sum_i [\log_e n_i - \log_e (g_i - n_i)]dn_i = 0 \qquad \ldots(v)$$

But $\sum_i dn_i = 0$ ...(vi)

and $\sum_i E_i dn_i = 0$ ...(vii)

Multiplying (vi) by α and (vii) by β and adding to equation (v), we get

$$\sum_i [\log_e n_i - \log_e (g_i - n_1) + \alpha + \beta E_i] dn_i = 0$$

The equilibrium distribution is possible if

$$\therefore \quad \log_e n_i - \log_e (g_i - n_1) + \alpha + \beta E_i = 0 \qquad \text{...(viii)}$$

$$\log_e \left(\frac{n_i}{g_i - n_i} \right) = -\alpha + \beta E_i$$

$$\frac{n_i}{g_i - n_i} = e^{-\alpha - \beta Ei}$$

$$\frac{g_i - n_i}{n_i} = e^{\alpha + \beta Ei}$$

$$\frac{g_i}{n_i} - 1 = e^{\alpha + \beta Ei}$$

$$\frac{g_i}{n_i} = (e^{\alpha + \beta Ei}) + 1$$

$$n_i = \left(\frac{g_i}{e^{\alpha + \beta E_i}} \right) + 1 \qquad \text{...(ix)}$$

Equation (ix) represents the Fermi-Dirac distribution law. The parameter has the same role as in the case of Maxwell-Boltzmann distribution law *i.e.*, for the system consisting of fermions in statistical equilibrium.

$$\beta = \frac{1}{kT}$$

$$\therefore \quad n_i = \frac{g_i}{[e^{(\alpha + E_i/kT)}] + 1}$$

In most cases the value of a is negative and is taken to be equal to

$$\frac{-E_F}{kT}$$

$$n_i = \frac{g_i}{\left[e^{(E_i - E_F)/kT)}\right] + 1}$$

The value of E_F is positive and is independent of temperature.

For T = 0, all the energy states are fully occupied and $n_i = g_i$ All the states with $E > E_F$, are empty *i.e.*, $n_i = 0$.

$$\left[e^{(E_i - E_F)/kT)}\right] = \begin{cases} 0 \text{ for } E_i - E_F < 0 \\ \infty \text{ for } E_i - E_F > 0 \end{cases}$$

In the case of Fermi-Dirac statistics, the accumulation of particles at the ground level is not allowed and at temperature T = 0, the particles occupy the lowest energy levels upto E_F. Here, the energy E_F gives the indication of the maximum energy of the fermions in the system. By is also called Fermi energy. For higher temperatures, the particles, occupy higher energy states greater than E_F. The curves indicate that only those fermions with energies close to E_F can move into unoccupied higher energy states. If $k\theta_F = E_F$, the temperature θ_F is called the Fermi temperature.

Maxwell-Boltzmann Distribution and Ideal Gas

Most of the gases obey Maxwell-Boltzmann distribution over a wide range of temperature. Consider a gas having mono-atomic molecules. It is assumed that the gas is ideal and has only kinetic energy of translation. Intermolecular attraction is assumed to be absent. If the kinetic energy is not quantized but considered to possess continuous values of energy, then the partition function Z can be written as

$$Z = \int_0^{\infty} e^{-(E/kT)}\, g(E)dE$$

Here g_i is replaced by g (E) dE

$$\text{But} \quad g(E)dE = \left(\frac{4\pi V(2m^3)^{1/2}}{h^3}\right)E^{1/2}dE$$

Here V is the volume occupied by the gas and h is Planck's constant.

$$Z = \frac{4pV(2m^3)^{1/2}}{h^3}\int_0^{\infty} E^{1/2}e^{-(E/kT)}dE$$

$$Z = \frac{4\pi V(2m^3)^{1/2}}{h^3}\left[\frac{\sqrt{\pi(kT)^3}}{2}\right]$$

$$Z = \frac{V(2\pi mkT)^{3/2}}{h^3} \qquad \text{...(i)}$$

Equation (i) represents the partition function for an ideal monoatomic gas in terms of the volume and temperature of the gas.

Taking logarithms on both sides

$$\log_e Z = \log_e \left[\frac{(2\pi m)^{3/2} V}{h^3}\right] + \log_e (kT)^{3/2}$$

$$\log_e Z = C + \frac{3}{2} \log_e kT \qquad \text{...(ii)}$$

Also $$E_{av} = (kT^2)\frac{d}{dT}(\log_e Z)$$

$$E_{av} = (kT^2)\frac{d}{dT}\left[C + \frac{3}{2}\log_e kT\right]$$

$$= (kT^2)\frac{d}{dT}\left[C + \frac{3}{2}\log_e k + \frac{3}{2}\log_e T\right]$$

$$E_{av} = \frac{3}{2}kT \qquad \text{...(iii)}$$

The total energy,

$$U = NE_{av}$$

$$U = N\left(\frac{3}{2}kT\right)$$

$$U = \frac{3}{2}NkT. \qquad \text{...(iv)}$$

Equation (iv) shows that the internal energy of an ideal monoatomic gas depends only on its temperature. The same relation does not however, hold good for real gases. In the case of real gases the internal energy is partly potential and partly kinetic. The total energy depends on the volume of the gas.

GIBBS FUNCTION

The Gibbs function G of a system is given by,

$$G = U - TS + PV$$

Consider a system that can do other forms of work, in addition to P.dV work, *e.g.*, a voltaric cell. In the case of a voltaic cell, the electrical work is –E.dI. Similarly for a magnetic material, the magnetic work is –m.dH. In general, the work will be given by P.dV plus a sum of terms each being the product of intensive variable (such as P, E or m) and the differential of an extensive variable (such as dV, dI and dH). In the case of a voltaic cell, the intensive variable is S and differential of extensive variable is dI. Suppose, in general, in addition to PdV the intensive variable is y and differential of extensive variable is dx. The work done for any reversible process,

$$\delta W = PdV + ydx$$

$$W = \int_{V_1}^{V_2} PdV + \int_{x_1}^{x2} ydx$$

Take $\int_{x_1}^{x2} ydx = A$

Consider a process where the system works at constant pressure P_0 and the change in volume is $(V_2 - V_1)$

$$\therefore \quad \int_{V_1}^{V_2} PdV = P_0 [V_2 - V_1]$$

$$\therefore \quad W = P_0[V_2 - V_1] + A \quad \text{...(i)}$$

For a system that exchanges heat with a reservoir temperature

$$W < (U_1 - U_2) - T_0 (S_1 - S_2) \quad \text{...(ii)}$$

$$P_0[V_2 - V_1] + A < (U_1 - U_2) - T_0 (S_1 - S_2)$$

$$\text{or} \quad A < (U_1 - U_2) - T_0 (S_1 - S_2) + P_0[V_1 - V_2] \quad \text{...(iii)}$$

Consider a specific process, where the initial and the final states of the system and the surroundings are at the same temperature (T_0) and pressure (P_0)

$$To = T \text{ and } P_0 = P$$

From equation (iii)

$$A_{P,T} < (U_1 - U_2)_{P,T} - T(S_1 - S_2)_{P,T} + P[V_1 - V_2]_{P,T} \quad \text{...(iv)}$$

But, for the Gibbs function,

$$G = U - TS + PV \quad \text{...(v)}$$

Therefore, for two equilibrium states at the same pressure and temperature,

$$[G_1 - G_2)_{P,T} = (U_1 - U_2)_{P,T} - T[S_1 - S_2]_{P,T} + P[V_1 - V_2]_{P,T} \quad \text{...(vi)}$$

From equations (iv) and (vi)

$$A_{P,T} < (G_1 - G_2)_{P,T} \quad \text{(vii)}$$

Thus, the difference between Gribbs function of a system between two equilibrium states sets the maximum limit to the work in addition to PdV work, provided the initial and the final states are at the same pressure and temperature and the system exchanges heat with a single heat reservoir. The work done will be maximum when the process is reversible. The process in this case will be isothermal–isobaric. If the process is irreversible, work done will be less than the maximum.

MAXWELL-BOLTZMANN DISTRIBUTION IN TERMS OF TEMPERATURE

The total energy of an isolated system is given by

$$U = \sum_i n_i E_i$$

$$U = n_1E_1 + n_2E_2 + n_3E_3 + ...$$

$$U = (g_1e^{-\alpha-\beta E1}) E_1 + (g_2e^{-\alpha-\beta E2}) E_2 + ...$$

$$U = e^{-\alpha}[g_1E_1e^{-\beta E1} + g_2E_2e^{-\beta E2} + ...]$$

But $$e^{-\alpha} = \frac{N}{Z}$$

$$\therefore \quad U = \frac{N}{Z}[g_1E_1e^{-\beta E1} + g_2E_2e^{-\beta E2} + ...]$$

$$U = \frac{N}{Z}\sum_i g_i E_i e^{-\beta Ei}$$

Here $$Z = \sum_i g_i e^{-\beta Ei} \qquad ...(i)$$

$$\frac{dZ}{d\beta} = \frac{d}{d\beta}\sum_i g_i e^{-\beta Ei} = -\sum_i g_i E_i e^{-\beta Ei}$$

$$\therefore \quad \sum_i g_i E_i e^{-\beta Ei} = \frac{d}{d\beta}\sum_i g_i E_i e^{-\beta Ei}$$

Substituting this value in equation (i)

$$U = -\frac{N}{Z}\frac{d}{d\beta}\sum_i g_i e^{-\beta Ei}$$

$$U = -\frac{N}{Z}\frac{d}{d\beta}$$

$$U = -N\frac{d}{d\beta}[\log_e Z] \qquad ...(ii)$$

The average energy of a particle is

$$E_{av} = -\frac{U}{N} = -\frac{d}{d\beta}(\log_e Z) \qquad ...(iii)$$

This shows that for a given system, the total energy U, the partition function Z, and the average energy of the particle E_{av}, depend on the parameter β.

Therefore, β may be taken to characterise the internal energy of the system and has the units per joule. If T is the temperature in degrees kelvin, then it is more convenient to represent parameter β as

$$\beta = \frac{1}{kT} \text{ or } KT = \frac{1}{\beta} \quad ...(iv)$$

Here kT has the unit of energy *i.e.*, joule and k is the Boltzmann's constant. Its units are J/K. The value of k is given by

$$k = 1.3805 \times 10^{-23} \text{ J/K}$$

Substituting the value of

$$\beta = \frac{1}{kT}$$

in all the equations, we get

$$Z = \sum_i g_i e^{-(Ei/kT)} \quad ..(v)$$

$$n_i = \frac{N}{Z} \sum_i g_i e^{-(Ei/kT)} \quad ...(vi)$$

This equation represents the Maxwell-Boltzmann distribution law in terms of the temperature of the system.

As $\beta = \dfrac{1}{kT}$

$$d\beta = -\frac{dT}{kT^2}$$

Substituting this value in equation (ii)

$$\therefore \quad U = (kNT^2)\frac{d}{dT}(\log_e Z) \quad ...(vii)$$

Also, from equation (iii)

$$E_{av} = (kT^2)\frac{d}{dT}(\log_e Z) \quad ...(viii)$$

Equation (viii) gives the relation between the average energy of the particle and its temperature under equilibrium position. Hence, the temperature of a system in statistical equilibrium is the physical quantity related to the average energy of the particle of the system.

ELECTRON GAS

Electrons in a metal belong to a most characteristic system of fermions because electrons obey the exclusion principle. For electrons in a metal, the energy levels are grouped in bands. Practically at all temperatures, the lower level energy bands are filled with electrons. The upper level energy bands are only partially filled with electrons. The distribution of electrons is to be considered only in the upper bands called the conduction band.

The zero energy level is taken at the lowest level of the conduction band. It is also assumed that the electrons have free movement within the conductor, provided the energy associated with the electrons is of the order of upper level energy bands.

As the energy of the electron in the conduction band is continuous, the term g_i is replaced by g(E) dE. Here dn electrons have energy in the range E and E + dE. According to Fermi-Dirac distribution law

$$n_i = \frac{g_i}{\left[e^{(E_i - E_F)/kT}\right] + 1} \qquad \text{...(i)}$$

Substituting the value of g_i = g (E) dB in equation (i) and replacing n_i by dn

$$dn = \frac{g(E)dE}{\left[e^{(E_i - E_F)/kT}\right] + 1}$$

As an electron has the spin ±1/2, the total number of states in the sphere is twice. $V/(2\pi)^3$ refers to the translational states per unit volume in the Fermi space Fermi sphere of radius kF has the total number of particles accommodated,

$$n = \frac{2\left[\frac{V}{(2\pi)^3}\right]\left(\frac{4}{3}\pi(kF)^3\right)}{\left[e^{(E_i - E_F)/kT}\right] + 1}$$

Here
$$E = \frac{p^2}{2m} \quad \frac{\left(\frac{h}{2\pi}\right)^2 (kF)^2}{2m}$$

$$\therefore \quad kF = \frac{p}{h/2\pi} = \frac{2\pi p}{h}$$

$$\therefore \qquad n = \frac{2\left[\frac{V}{(2\pi)^3}\right]\left(\frac{4}{3}\pi\left(\frac{2\pi p}{h}\right)^3\right)}{\left[e^{(E_i - E_F)/kT)}\right]+1}$$

But $p = (2mE)^{1/2}$

$$\therefore \qquad n = \frac{2\left[\frac{V}{(2\pi)^3}\right]\left[\left(\frac{4}{3}\pi\left(\frac{2\pi p}{h}\right)^3 (2mE)^{3/2}\right)\right]}{\left[e^{(E_i - E_F)/kT)}\right]+1}$$

$$n = \frac{\left(\frac{8\pi V}{(3h^3}\right)(2m)((2m)^{1/2} E^{3/2}}{\left[e^{(E_i - E_F)/kT)}\right]+1}$$

Differentiating

$$dn = \frac{\left(\frac{8\pi V}{(3h^3}\right)(2m)((2m)^{1/2} \times \frac{3}{2} E^{1/2} dE}{\left[e^{(E_i - E_F)/kT)}\right]+1}$$

$$\frac{dn}{dE} = \frac{\left[\frac{8\pi V(2m^3)^{1/2}}{(3h^3}\right]E^{1/2}}{\left[e^{(E_i - E_F)/kT)}\right]+1} \qquad \text{...(ii)}$$

Equation (ii) represents the energy distribution for free electrons. This is also called Fermi-Dirac formula of free.fermions.

At $\qquad T = 0K,$

$$p = (2m\,E_F)^{1/2} \qquad \text{...(iii)}$$

and $N = \left(\frac{2V}{h^3}\right)\left(\frac{4}{3}\pi p^3\right)$

or
$$P = \left(\frac{3Nh^3}{8\pi V}\right)^{1/3} \qquad \text{...(iv)}$$

Equating (iii) and (iv)

$$(2mE_F)^{1/2} = \left(\frac{3Nh^3}{8\pi V}\right)^{1/3}$$

Squaring $2mE_F)^{1/2} = \left(\frac{3Nh^3}{8\pi V}\right)^{2/3}$

or $$E_F = \frac{h^2}{8m}\left(\frac{3N}{\pi V}\right)^{2/3} \qquad ...(v)$$

Knowing the value of N/V *i.e.*, the number of free electrons per unit volume, the Fermi energy for electrons in a metal can be obtained.

For silver, the value $\frac{N}{V}$

$$= 5.86 \times 10^{28} \text{ electrons/m}^3$$

$$\therefore \quad E_F = \left(\frac{(6.624\times10^{-34})^2}{8\times9\times10^{-31}}\right)\left(\frac{3\times5.86\times10^{28}}{3.14}\right)^{2/3}$$

$$E_F = 9 \times 10^{-19} J.$$

But $1eV = 1.6 \times 10^{-19}J$

$$\therefore \quad E_F = \frac{9\times10^{-19}}{1.6\times10^{-19}}$$

$$E_F = 5.625 \text{ eV}.$$

It means that the maximum kinetic energy of free electrons in silver at absolute zero temperature is 5.6 eV.

The Fermi temperature θ_F is given by

$$\theta_F = \frac{E_F}{k}$$

As E_F practically independent of temperature, the value θ_F is fixed for a given metal. The values of θ_F and E_F are given in the following table.

Metals	E_F ***(electron Vol.ts)***	θ_F***(K)***
Potassium	2.14	2.4×10^4
Sodium	312	3.7×10^4
Lithium	4.72	6.5×10^4
Silver	5.51	6.4×10^4
Gold	5.64	6.4×10^4
Copper	7.04	8.2×10^4

DISTRIBUTION LAW OF BOSE-EINSTEIN

Bose-Einstein distribution is applied to systems composed of identical and indistinguishable particles that are not restricted by the exclusion principle. In such systems, there is no limit to the number of particles occupying a particular quantum state. These particles are called *bosons*. Their spin is zero or 1. Mesons and helium nuclei are examples of bosons.

The Bose-Einstein statistics values of g_i refer to degeneracy of each energy level. Suppose, that n_i particles are arranged in a row and distributed among g_i quantum states with $(g_i - 1)$ partitions in between.

The total number of possible arrangements of particles and partitions is equal to the total number of permutations of $(n_i + g_i - 1)$ objects in a row. Therefore the total possible ways of arranging n_i particles with $g_i - 1$ partitions

$$= (n_i + g_i - 1)!$$

As the particles are identical and indistinguishable the possible number of distinct arrangements

$$= \frac{(n_i + g_i - 1)!}{n_i!(g_i - 1)!}$$

The total number of distinguishable and distinct ways of arranging N particles in all the available energy states is given by

$$P = \frac{(n_1 + g_1 - 1)!}{n_1!(g_1 - 1)!} \times \frac{(n_2 + g_2 - 1)!}{n_2!(g_2 - 1)!} \times \ldots$$

$$P = \prod_i \frac{(n_i + g_i - 1)!}{n_i!(g_i - 1)!} \qquad \ldots(i)$$

The most probable distribution can be obtained by finding the maximum value of $\log_e$, P.

According to Stirling's approximation

$$\log_e x! = x \log_e x - x$$

Also $$\sum_i n_i = N \qquad \ldots(ii)$$

and $$\sum_i n_i E_i = U \qquad \ldots(iii)$$

From equation (i), applying Stirling's approximation

$$\log_e P = \sum_i [\log_e (n_i + g_i - 1) - \log_e n_i! \log_e (g_i - 1)!]$$

$$\log_e P = \sum_i [(n_i + g_i - 1)\log_e(n_i + g_i - 1) - (n_i + g_i - 1)$$

$$- (n_i \log_e n_i - n_i) - (g_i - 1)\log_e(g_i - 1) - (g_i - 1)]$$

$$\log_e P = \sum_i [(n_i + g_i - 1)\log_e(n_i + g_i - 1)$$

$$- n_i \log_e - n_i - (g_i - 1)\log_e(g_i - 1)] \quad ...(iv)$$

The maximum value of P is obtained by taking

$d(\log_e P) = 0.$

Differentiating equation P.

$$d(\log_e P) = \sum_i [\log_e(n_i + g_i - 1)\, dn_i - \log_e n_i - dn_i] = 0$$

$$\therefore \quad -d(\log_e P) = \sum_i [-\log_e(n_i + g_i - 1)\, dn_i + \log_e n_i dn_i] = 0$$

or

$$\sum_i [-\log_e(n_i + g_i - 1) + \log_e n_i] dn_i = 0 \quad ...(v)$$

As the total number of particles and total energy are constants, we have,

$$\sum_i dn_i = 0 \quad ...(vi)$$

$$\sum_i E_i dn_i = 0 \quad ...(vii)$$

Multiplying (vi) by α and equation (vii) by β and adding to equation (v), we get

$$\sum_i [-\log_e(n_i + g_i - 1) + \log_e n_i + \alpha + \beta E_i] dn_i = 0$$

$$\therefore \quad -\log_e(n_i + g_i - 1) + \log_e n_i + \alpha + \beta E_i = 0$$

Taking $n_i + g_i$ very large as compared to 1, the quantity 1 can be neglected.

$$\therefore \quad -\log_e(n_i + g_i) + \log_e n_i + \alpha + \beta E_i = 0$$

$$\log_e\left(\frac{n_i}{ng_i + g_i}\right) = -\alpha - \beta E_i$$

$$\frac{n_i}{n_i + g_i} = e^{-\alpha - \beta Ei}$$

$$\frac{n_i + g_i}{n_i} = e^{\alpha + \beta Ei}$$

$$1 + \frac{g_i}{n_i} = e^{\alpha + \beta Ei}$$

$$\frac{g_i}{n_i} = (e^{\alpha + \beta Ei}) - 1$$

$$n_i = \left(\frac{g_i}{e^{\alpha + \beta E_i}}\right) - 1 \qquad \text{...(viii)}$$

Equation (ix) represents the Bose-Einstein distribution law.

Taking $\beta = \frac{1}{kT}$

$$\therefore \qquad n_i = \frac{g_i}{[e^{(\alpha + E_i/kT)}] - 1}$$

The value of the constant a is governed by the equation $\Sigma n_i = N$. As n_i cannot be negative, α must always have a positive value. The distribution of the particles for different energy levels

STATISTICS APPLIED TO MOLECULES

A monatomic perfect gas. The state of each molecule of the gas in the μ-space is represented by a point (and by a cell in quantum mechanical terms). But, since the gas molecules are in constant motion, their state will be changing continuously in space and, due to collisions (and also mutual interactions), there is a change .in momentum state also. The trajectory of such a molecule constant energy surface given by

$$E = \frac{p_x^2 + p_y^2 + p_z^2}{2m} = V(q_x, q_y, q_z)$$

The real gas molecules, however, are endowed with mutual interactions in addition to usual collisions, hence their state is better represented by a point in y-space. The rate of change of its state, then, depends on the quantum of interactions. The path of representative point in y-space is given by the relation

$$E = \sum_i T_i + \sum_i V_i + \sum_{i,j}' U_{ij}$$

where T_i, V_i and U_{ij} refer to kinetic energy, potential energy and energy of interaction of ith particle. The prime on the summation sign suggests that i = j is not allowed (which would mean self-interaction). With this background and remembering that the number of molecules is very large in a macrosystem, we now proceed to investigate the distribution details.

STEFAN'S LAW VERIFICATION

In 1897, Lummer and Pringsheim experimentally verified Stefan's law over a wide range of temperature (100°C to 1,300°C). The apparatus consisted of a black body C. For temperatures between 200°C and 600°C, a hollow copper sphere coated inside with platinum black was used. The fused nitrates of sodium and potassium having a melting point of 219°C were used as the bath surrounding the black body. For temperature between 900°C and 1,300°C, an iron cylinder coated inside with platinum black was used as a black body and it was enclosed in a double walled gas furnace. A thermocouple T was used as a thermometer. A bolometer B was used to measure the intensity of the emitted heat radiations. S_1, S_2 and S_3 were the water-cooling shutters. Another black body A at 100°C was used to standardize the bolometer.

The double walled vessel of the black body A contained boiling water at 100°C. The bolometer B was allowed to face the opening of the black body A and the shutter S_3 raised. The deflections in the galvanometer of the bolometer at various distances were noted and it was found that the deflection was inversely proportional to the square of the distance between the bolometer and the opening of the black body A. Thus, the deflection in the galvanometer was proportional to the intensity of heat radiations.

The shutter S_3 was closed and the bolometer B was allowed to face the opening of the black body C. The shutters of S_1 and S_2 were raised. The bath surrounding the black body was maintained at a constant temperature and the maximum deflection produced in the galvanometer of the bolometer was noted. Thus at various constant temperatures of the black body, corresponding to constant deflections (in the galvanometer of the bolometer) were observed. Then the data was reduced to a common arbitrary unit in terms of the total radiations from the black body A at 100°C.

Let θ be the deflection in the galvanometer, T_1 the temperature of the black body and T_2, the temperature at the entrance of the bolometer. It was found that

$$\theta \propto (T_1^4 - T_2^4)$$

But $\theta \propto R$

$$\therefore \qquad \theta \propto (T_1^4 - T_2^4)$$

This verifies Stefan's law.

Recently Coblentz has verified Stefan's law more accurately. He took an electrically heated black body whose temperature was measured by an accurate thermocouple. An absolute bolometer was used to measure the amount of heat radiations emitted by the black body. He was able to show the correctness of Stefan's law experimentally up to 1,600°C.

Macroscopic Systems Characteristic Features

Let us now discuss some of the characteristic features of the macrosystems. In doing so we will assume that it has a large number of microscopic particles.

Fluctuation

Let the isolated system has N molecules of a perfect gas and that it was left to itself for quite some time so that it is in equilibrium. Further, also consider that the system has been partitioned into two identical (also equal) parts by an imaginary wall, each part A and B has n and n molecules respectively such that n + n = N. Now, as we have seen, for large values of N, $n \cong n^1 = N/2$

But the molecules of the gas are in a state of constant random motion. So that some of the molecules are leaving portion .A and moving into B and vice-versa. Thus, there is a continuous change in the number of molecules in each portion of the box. Time variation of this change is known as fluctuation. The state of normal fluctuation of an isolated system. Usually, N/2 – n is only a small number. For large deviation, the number of microstates (and hence the probability) reduce drastically. The number of microstates for different values of N and N/2 – n. For the sake of simplicity the deviation N/2– n is represented as a fraction of total number of particles. Value of R ($\cong P_x/P_{max}$) is also shown to arrive at qualitative results; In actual physical problems the number of molecules are $\sim 10^{24} m^{-3}$ whence even 0.01% deviation also becomes quite improbable. This is the reason for quick diffusion and most equitable distribution of air molecules in the available space.

Theorems of Probability in Statistical Thermodynamics

The following are the important probability theorems commonly used in statistical thermodynamics:

(1) The number of ways in which N distinguishable particles can he aıranged in order is equal to

$$N!$$

(2) The number of different ways in which N particles can be selected from N distinguishable particles irrespective of the order of selection is equal to

$$\frac{N!}{(N-n)!n!}$$

(3) The number of different ways in which n indistinguishable particles can be arranged in g distinguishable states with not more then one particle in each state is equal to

$$\frac{g!}{(g-n)!n!}$$

PROBABILITIES CALCULATION

We can extend the theory of probability in order to include the occurrence of two or more independent events. The combined probability in such cases is given by the product of separate probability of each event. Let us examine some cases:

(i) *Tossing of a coin:* On tossing a single coin, there are two possible aspects that may be presented by a fallen coin, a "head" or a "tail". We may assume here that there is no preference for any particular event, so that after making a large number of throws, the ratio of the number of heads to the number of tails is unity. Thus, for a normal coin the probability of each event is 1/2.

(ii) *Tossing of two unlike coins*: Suppose we toss two coins of different denominations or of different sizes, or of different colours, we have four complexions.

1. Head_1 Head_2
2. Head_1 Tail_2
3. Tail_1, Tail_2
4. Tail_1 Head_2.

Thus, the probability of any one complexion on a single toss will be 1/4 = 1/2.1/2, Or the composite probability is the product of probability of individual events.

(iii) *Tossing of two similar coins*: If two coins cannot be distinguished from each other, then complexions 2 and 4 of (ii) are indistinguishable whence the probability of one head and one tail is twice than that of two heads or two tails.

(iv) *Tossing of two coins in general*: Let us represent occurrence of heads by A_1, A_2 and the occurrence of tails by B_1, B_2 when two coins are tossed. Then the composite event can be represented by $(A_1 + B_1)(A_2 + B_2) = A_1A_1 + A_1B_2 + A_2B_1 + B_1B_2$.

(v) *Tossing of three unlike coins:* With three unlike coins, we have three distinguishable heads A_1, A_2, A_3 and three distinguishable tails, B_1, B_2, B_3. Here we can have eight equally probable complexions $A_1 A_2 A_3$. $A_1 A_2 B_3$, $A_1 B_2 B_3$, $B_1 A_2 B_3$, $B_1 A_2 B_3$, $B_1 B_2 A_3$ and $B_1 B_2 B_3$ which are really obtained from the product of $(A_1 + B_1)$, $(A_2 + B_2)$, $(A_3 + B_3)$.

(vi) *Tossing of three similar coins:* In this case we have only four probable states with complexions

1. All heads.
2. Two heads and one tail
3. All tails.
4. One head and two tails.

We can tabulate the result as in (iv) for

$(A + B)(A + B)(A + B) = A_3 + 3A^2B + 3AB^2 + B^3$.

(vii) *Tossing of n coins*: Suppose all the coins are different. Then the possible complexions (also known as probability number) will be $(A_1 + B_1)(A_2 + B_2) \ldots (A_n + B_n)$. And, if the coins are all alike the total number of possible complexions will be

$$(A + B)^n = A^n + {}^nC_1A^{n-1}B + {}^nC_2A^{n-2}B^2 + \ldots$$
$$\qquad {}^nC_rA^{n-r}B^r + \ldots + {}^nC_n B^n$$

$$\text{or} \qquad = \sum_{r=0}^{r=n} C_r A^{n-r}B^r$$

$$\text{or, also } (A + B)^n = \sum_{r=0}^{r=n} C_r A^rB^{n-r}$$

$$\equiv \sum_{r=0}^{r=n} C_r A^{n-r}B^r$$

where $^nC_r = \frac{n!}{r!(n-r)!}$ ≡ Number of events with r heads/tails.

Now, since A and B indicate one event each, the total number of events are $(1 + 1)^n = 2^n = {}^nC_1 + {}^nC_2 + ... + {}^nC_n$

∴ The probability of r heads and (n – r) tails to appear is

$$P(r,n-r) = \frac{^nC_r}{2^n} = \frac{n!}{r!(n-r)!} \cdot \frac{1}{2^n}$$

Maximum and minimum probability: To find out which of the combination of heads and tails is most likely in a large number of throws of n coins, we have to find the maximum value of ^nC_r. It is known from algebra that the value of ^nC_r is greatest for r = n/2 when n is even and r = n+1/2 when n is odd.

$$P_{max} = \frac{^nC_{n/2}}{2^n} = \frac{n!}{(n/2)!(n/2)} \cdot \frac{1}{2^n}$$

Similarly, the minimum probability, P_{min} obtained for r = 0 or r = n as $^nC_n = 1$.

whence $P_{min} = 1/2^n$

(viii) *Tossing of a single coin n times*: When we toss the same coin second, third,... nth time, its behaviour is again unpredictable. It may show a head or a tail each time with 1/2 probability. Hence a single coin tossed n times (and of course, making large number of such attempts, if we want to have satisfactory resuits) is just equivalent to the case of tossing of n coins as discussed in (vii).

Another equivalent illustration can be: Take 2 boxes, identical in all respects, and throw n balls from a great distance. Assuming that each ball will fall into one or the other box, the distribution of balls in the boxes will be governed by the relation (vii),

(ix) *Tossing of a dice*: If we toss of dice, having six faces, the probability for each face showing up will be 1/6. Since now the total number of aspects is 6 and only one can show up at a time with equal probability. Again, we may take 2, 3,.... n dices and throw them and calculate the composite probability for a certain complexion.

(x) *Tossing of n dices each having m faces*: Here let us assume that dice 1 has faces $a_1, a_2, a_3... a_m$; dice 2 has faces $b_1\ b_2\ b_3\ ...\ b_m$;

dice 3 has faces c_1, c_2, c_3 ; etc., then the possible complexions with n dices will be the sum of each aspect

$a_1, a_2, a_3 ... a_m, v_1, b_2, b_3 ... b_m, c_1, c_2, c_3$ ... n sets and, if the dice are all alike, we have

$m_{.a}, m_{.b}, m_{.c}$... n sets.

If the probability for occurring each of the face is same, *i.e.*, a = b = c... = one aspect, we get total number of possible aspects

= m.m.m.... n times = m^n

One can show that the most probable distribution corresponds to equitable distribution of faces *i.e.*, face 1, face 2, face 3, etc., will appear in equal number. Then all the dices showing face 1 have the probability of only $1/m^n$.

A similar illustration to this is : Take *m identical boxes and throw n* balls onto them from a great distance. We will take up this problem later in more detail. As in actual analysis of physical problems we have a large number of particles which are distributed into a large number of (hypothetical) boxes.

(xi) *Probability with weightage*: So far we have considered the case in which the probability of occurring an event and its failure is the same. But there can be cases where the probability of occurring a particular event is greater than that for any other event.

Let a be the probability of occurring an event and b be the probability of its failure, per trial. Then, if the trial is repeated n times, the probability of r favourable events is

$a^r.b^{n-r} = a^r (1 - a)^{n-r}$

But there are ^nC_r ways for r favourable events, therefore, the probability of occurring r events is

$$P(r, n - r) = {^nC_r}\, a^r (1 - a)^{n-r}$$

Thus, for example, in a dice a = 1/6

$$\therefore \quad P(r, n - r) = {^nC_r} \left(\frac{1}{6}\right)^r \left(\frac{5}{6}\right)^{n-r}$$

But, if certain number, say 3, has a = 2/7 then in n tosses the number 3 will occur r times with a probability.

(xii) *Postulate of equal a priori probability:* It states that all accessible microstates corresponding to a macrostate are equally probable.

Thus, the postulate claims that the probability of finding a molecule in one region of phase space is identical with that for any other region of equal volume provided the regions are obeying other conditions of equality (*i.e.*, the regions have extensions of same magnitude and correspond to the same energy). Thus, it follows that the probability of occurrence of a macrostate is proportional to the number of microstates associated with it

i.e., $P_i \propto \Omega_i$

where P_i is the probability and Ω_i is the total number of accessible microstates, in the ith set.

SOME SOLVED PROBLEMS

Problem 1:

(i) Calculate α *and* β *for an ideal gas obeying gas equation PV = RT (for one mole).*

Solution:

(i) We know

$$PV = RT$$

On differentiating the above equation, we obtain

$$PdV + VdP = RdT \qquad ...(1)$$

At constant pressure dP = 0, the above equation becomes as follows:

$$\left(\frac{\partial V}{\partial T}\right)_P = \frac{R}{V}$$

or

$$\alpha = \frac{1}{V}\left(\frac{\partial V}{\partial T}\right)_P = \frac{R}{VP} = \frac{1}{T} = T^{-1}$$

At constant temperature dT = 0; equation (1) becomes as follows:

$$\left(\frac{\partial V}{\partial P}\right)_T = -\frac{V}{P}$$

or

$$\beta = -\frac{1}{V}\left(\frac{\partial V}{\partial P}\right)_T = \frac{V}{VP} = \frac{1}{P} = P^{-1}$$

(ii) The van der Waal's equation is as follows :

$$\left(P + \frac{a}{V^2}\right)(v - b) = RT$$

or $$PV^3 - PbV^2 + aV - ab - RTV^2 = 0 \qquad ...(2)$$

On differentiating the above equation with respect to T at constant pressure, we obtain

$$3\,PV^2\left(\frac{\partial V}{\partial T}\right)_P - 2PbV\left(\frac{\partial V}{\partial T}\right)_P + a\left(\frac{\partial V}{\partial T}\right)_P$$

$$- 2RTV\left(\frac{\partial V}{\partial T}\right)_P - RV^2 = 0$$

or $$\left(\frac{\partial V}{\partial T}\right)_P = \frac{RV^2}{3PV^2 - 2PbV + a - 2RTV}$$

$$\alpha = \frac{1}{V}\left(\frac{\partial V}{\partial T}\right)_P$$

$$= \frac{1}{V}\left[\frac{RV^2}{3PV^2 - 2PVb + a - 2RTV}\right]$$

$$= \frac{R}{3PV - 2Pb + \frac{a}{V} - 2RT}$$

But the van der Waal's equation may be put as follows :

$$RT = PV - Pb + \frac{a}{V} - \frac{ab}{V^2}$$

$$\therefore \qquad \alpha = \frac{R}{3PV - 2Pb + \frac{a}{V} - 2\left(PV - Pb + \frac{a}{V} - \frac{ab}{V^2}\right)}$$

$$= R\left[PV - \frac{a}{V} + \frac{2ab}{V^2}\right]^{-1}$$

Again, on differentiating equation (2), with respect to P at constant temperature T, we obtain

$$3PV^2\left(\frac{\partial V}{\partial P}\right)_T + V - 2PbV\left(\frac{\partial V}{\partial P}\right)_T$$

$$- bV^2 + a\left(\frac{\partial V}{\partial P}\right)_T - 2RTV\left(\frac{\partial V}{\partial P}\right)_T = 0$$

$$\left(\frac{\partial V}{\partial P}\right)_T = \frac{V^2(b - V)}{3PV^2 - 2PbV + a - 2RTV}$$

But the van der Waal's equation may be put as follows :

$$RTV = PV^2 + a - PbV - ab/V$$

$$\therefore \quad \left(\frac{\partial V}{\partial P}\right)_T = \frac{V^2 (b - V)}{3PV^2 - 2PbV + a - 2\left(PV^2 + a - PbV - \frac{ab}{V}\right)}$$

$$= \frac{b - V}{P - \frac{a}{V^2} + \frac{2ab}{V^3}}$$

Now $\quad \beta = -\frac{1}{V}\left(\frac{\partial V}{\partial P}\right)_T = -\frac{b - V}{PV - \frac{a}{V} + \frac{2ab}{V^2}}$

$$= -\frac{V - b}{P - \frac{a}{V} + \frac{2ab}{V^2}}$$

Problem 2:

Derive $\quad C_{P,m} - C_{V,m} = TV_m \dfrac{\alpha^2}{\beta}$

where $\propto$ *and* β *are the coefficient of thermal expansion and coefficient of compressibility, respectively.*

Solution:

We start with H = f (T, P) and U = f (T, V) and write their differentials as

$$dH = \left(\frac{\partial H}{\partial T}\right)_P dT + \left(\frac{\partial H}{\partial P}\right)_T dP$$

$$dU = \left(\frac{\partial U}{\partial T}\right)_V dT + \left(\frac{\partial U}{\partial V}\right)_T dV$$

Now since $\quad H = U + PV$

we have $\quad dH = dU + P\,dV + V\,dP$

or $\quad \left(\frac{\partial H}{\partial T}\right)_P dT + \left(\frac{\partial H}{\partial P}\right)_T dP = \left(\frac{\partial U}{\partial T}\right)_V dT + \left(\frac{\partial U}{\partial V}\right)_T dV + P\,dV + V\,dP$

Dividing by dT, keeping P constant, we get

$$\left(\frac{\partial H}{\partial T}\right)_P = \left(\frac{\partial U}{\partial T}\right)_V + \left(\frac{\partial U}{\partial V}\right)_T \left(\frac{\partial V}{\partial T}\right)_P + P\left(\frac{\partial V}{\partial T}\right)_P$$

Now using the thermodynamic equation of state

$$\left(\frac{\partial U}{\partial V}\right)_T = T\left(\frac{\partial P}{\partial T}\right)_V - P$$

we get
$$\left(\frac{\partial H}{\partial T}\right)_P = \left(\frac{\partial U}{\partial T}\right)_V + T\left(\frac{\partial P}{\partial T}\right)_V\left(\frac{\partial V}{\partial T}\right)_P \quad ...(1)$$

Now, by definition

$$\alpha = \frac{1}{V}\left(\frac{\partial V}{\partial T}\right)_P \quad ...(2)$$

$$\beta = -\frac{1}{V}\left(\frac{\partial V}{\partial P}\right)_T \quad ...(3)$$

Using the cyclic rule

$$\left(\frac{\partial V}{\partial P}\right)_T\left(\frac{\partial P}{\partial T}\right)_V\left(\frac{\partial T}{\partial V}\right)_P + 1 = 0$$

we get
$$\left(\frac{\partial V}{\partial P}\right)_T = -\left(\frac{\partial V}{\partial T}\right)_P\left(\frac{\partial T}{\partial P}\right)_V$$

Hence

$$\beta = \frac{1}{V}\left(\frac{\partial V}{\partial T}\right)_P \cdot \left(\frac{\partial T}{\partial P}\right)_V$$

and
$$\frac{\alpha}{\beta} = \left(\frac{\partial P}{\partial T}\right)_V$$

Hence, Eq (1) becomes

$$C_{P,m} - C_{V,m} = T\left(\frac{\partial P}{\partial T}\right)_V\left(\frac{\partial V}{\partial T}\right)_P \quad ...(4)$$

With the use of Eqs. (2) and (4), we get

$$C_{P,m} - C_{V,m} = TV\frac{\alpha^2}{\beta}$$

Problem 3:

When a bicycle tyre is inflated with hand pump, the temperature of the air inside increases. Explain.

Solution:

Work done on the system is converted into heat which increases the temperature of air inside the tyre.

Problem 4:

Can q become a state function ?

Solution:

Heat exchange reversibly at constant volume (or at constant pressure) is equal to the change in internal energy (or enthalpy) of a system. Since the latter is a state function, it follows that in such conditions q also behaves like a state function.

Problem 5:

One mole of hydrogen and nine moles of nitrogen are mixed at 291. K and I atm pressure. Assuming ideal behaviour for the gases, calculate the entropy of mixing per mole of the mixture formed. Would it make any difference if under similar conditions one mole of hydrogen is mixed with nine moles of oxygen.

Solution:

For ideal gases the entropy of mixing per mole of the mixture is given as follows

$$\Delta S_{mtxing} = -R \sum X_i \ln X_i$$

In this case $X_{H2} = 0.1$ and $X_{N2} = 0.9$

$$\Delta S mixing = -R\,[0.1 \ln 0.1 + 0.9 \ln 0.9]$$

$$= -(8.314\ JK^{-1}\ mole^{-1})\,(2.303)\,(0.1)$$

$$[\because R = 8.314\ JK^{-1}\ mol^{-1}]$$

$$\left[\log\frac{1}{10} + 9\log\frac{9}{10}\right]$$

$$= 2.704\ \ JK^{-1}\ mol^{-1}$$

When hydrogen is mixed with oxygen, the entropy of mixing would remain unchanged provided oxygen also behaves ideally.

Problem 6:

2 moles of an ideal monoatomic gas ($C_V = 20.91\ J\ mol^{-1}\ K^{-1}$) are heated from 700 K to 750 K. The volume of the gas changes from 0.08 m^3 to 0.8 m^3. Assuming that the heat capacity remains constant in this temperature range, calculate the entropy change for the system, the surroundings and the universe if the process is carried out (a) reversibly *and (b) irreversibly by placing the system in contact with a reservoir at 500 K and allowing the gas to expand against a const. external pressure equal to the final pressure of the gas*

Solution:

For the reversible transformation of an ideal gas the entropy change of the system is given as follows :

$$\Delta S_{system} = nC_V \ln \frac{T_2}{T_1} + nR \ln \frac{V_2}{V_1}$$

$$= (2\ \text{mol})\ (20.9\ \text{J mol}^{-1}\ \text{K}^{-1})\ (2.303)\ \log \left(\frac{750}{700}\right)$$

$$+ (2\ \text{mol})\ (8.314\ \text{J mol}^{-1}\ \text{K}^{-1})\ (\ 2.303)\ \log \frac{0.80}{0.08}$$

$$= 6.449 + 38.28 = 44.729\ \text{JK}^{-1}.$$

As the process is reversible, the entropy change of the surrounding is as follows.

$$\Delta S_{sur} = -\ 44.729\ \ \text{JK}^{-1}$$

and the entropy change of the universe is as follows

$$\Delta S_{univ} = \Delta S_{sys} + \Delta S_{surr} = 0$$

(b) In the irreversible transformation the entropy change of the system will be the same. Heat lost by the surrounding is as

$$-\,Q = \Delta E - W$$

$$\Delta E = -\ n\ C_V\,(750 - 700)$$

and $$\omega = P_{ext}\ (V_2 - V_1).$$

Thus according to the final state

$$T_2 = 750\ \text{K}$$

$$V_2 = 0.8\ \text{m}^3$$

and hence the pressure is given at follows

$$P_2 = \frac{nRT_2}{V_2} = \frac{(2\ \text{mol})\left(8.314\ \text{J mol}^{-1}\ \text{K}^{-1}\right)(750\,\text{K})}{\left(0.8\ \text{m}^3\right)}$$

$$= 15588.75\ \text{Jm}^{-3}.$$

Now $P_{ext} = P_2 = 15588.75\ \ \text{Jm}^{-3}$ and hence the heat lost by the surrounding is

$$-Q = -(2 \text{ mol})(20.9 \text{ J mol}^{-1} \text{K}^{-1})(50 \text{ K})$$

$$-(15588.75 \text{ Jm}^{-3})(0.8 \text{ m}^3 - 0.08 \text{ m}^3)$$

$$= -2090 \text{ j} - 11223.9$$

$$= 13313.9 \text{ J.}$$

Let us assume that this heat is lost reversibly by the surroundings. Its entropy change is given as

$$\Delta S_{surr} = \frac{13313.9 \text{ j}}{500 \text{ K}} = 26.6278 \text{JK}^{-1}$$

and the entropy change of the universe is

$$\Delta S_{univ} = \Delta S_{sys} + \Delta S_{surr}$$

$$= (44.729 - 26.6278) \text{ JK}^{-1}$$

$$= 18.1012 \text{ JK}^{-1}.$$

Problem 7:

The temperature of an ideal monoatomic gas is increased from 546 K to 1638 K. Calculate the pressure change in order that the entropy of the gas remains unchanged in the process

Solution:

The entropy change for the system can be explained by the following relation which is as follows

$$\Delta S_{sys} = nC_P \ln \frac{T_2}{T_1} \text{ nR} \ln \frac{P_2}{P_1}$$

$$\Delta S_{sys} = 0, \text{ thus}$$

$$C_P \ln \frac{T_2}{T_1} = R \ln \frac{P_2}{P_1}$$

$$\frac{5}{2} R \log \frac{T_2}{T_1} = R \log \frac{P_2}{P_1}$$

$$\frac{5}{2} \log \frac{1638 \text{ K}}{546 \text{ K}} = \log \frac{P_2}{P_1}$$

$$\frac{P_2}{P_1} = 15.60$$

Thus in order to make entropy constant, the pressure of the gas must be increased by 15.60 times its initial value.

Problem 8:

Two moles of an ideal gas are allowed to expand isothermally from $0.04\ m^3$ to $0.4\ m^3$ at 300 K. Calculate the entropy change for the system, surroundings and universe if the expansion is

(i) reversible and

(ii) irreversible against a constant external pressure of 0.2 atm.

Solution:

$$Q_{rev} = W = n\ RT\ \ln \frac{V_2}{V_1}$$

$$\Delta S_{rev} = \frac{Q_{rev}}{T} = nR\ \ln \frac{V_2}{V_1}$$

$$= (2\ mol)\ (8.314)\ J\ mol^{-1}\ K^{-1}\ (2.303)\ \log \frac{0.40}{0.04}$$

$$= 38.294284\ JK^{-1}.$$

The surroundings is losing an exactly equivalent amount of heat and hence its entropy decrease is given as follows :

$$\Delta S_{surr} = \frac{(-Q_{rev})}{T} = -19.14\ JK^{-1}$$

The entropy change of the universe is given as follows :

$$\Delta S_{univ} = \Delta S_{sys} + \Delta S_{surr} = 0$$

(ii) In the irreversible expansion, the entropy change of the system is still the same as in the reversible change, *i.e.*, 19.14 JK^{-1} but heat lost by the surroundings is given by $-P\Delta V$ and consequently the entropy change, if the transfer were reversible, is

$$\Delta S_{surr} = \frac{Q_{surr}}{T} = -\frac{P\Delta V}{T}$$

$$= -\frac{(0.2\,atm.)\left(0.4 - 0.04\,m^3\right)\left(101325\ Jm^{-3}\ atm^{-1}\right)}{300\ K}$$

$$= -\frac{0.2 \times 0.36 \times 101325}{300}\ JK^{-1}$$

$$= 24.318\ JK^{-1}..$$

Problem 9:

Calculate the entropy change when 1 mole of perfect gas is allowed to expand at 27°C from a volume of 2 litres to a volume of 20 litres against a constant pressure of one atmosphere.

Solution:

We know

$$dS = C_V \frac{dT}{T} + R \frac{dV}{V}$$

By integrating this equation, we get

$$S_2 - S_1 = C_V \ln \frac{T_2}{T_1} + R \ln \frac{V_2}{V_1}$$

At constant temperature,

$$S_2 - S_1 = R \ln \frac{V_2}{V_1}$$

$$= 2.303 R \log \frac{V_2}{V_1}$$

$$= 2.303 \times 2 \times \log \frac{20}{2}$$

$$= 2.303 \times 2 \times \log 10$$

= 4.606 cal./ degree/mole.

$$\left[\begin{array}{l} \because R = 2 \text{ cal./degree/mole} \\ V_1 = 2 \text{ litres} \\ V_2 = 20 \text{ litres} \end{array}\right]$$

Problem 10:

The molar heat capacity at capacity at constant pressure of solid magnesium (0° – 600°C) is expressed by

CP = 6.20 + 1.30 × 10–3 T – 6.80 × 104 T–2

Calculate the increase in entropy when 1 gm atom of metal is heated from 27°C to 127°C at constant pressure.

Solution:

At constant pressure,

$$\Delta S = \int_{T_1}^{T_2} \frac{C_P}{T}\, dT$$

$$\therefore \quad \Delta S = \int_{300}^{400} \frac{\left(6.20 + 1.30 \times 10^{-3}\,T - 6.80 \times 10^{4}\,T^{-2}\right) dT}{T}$$

$$\int_{300}^{400} 6.20 \frac{dT}{T} + \int_{300}^{400} 1.3 \times 10^{-3} dT - \int_{300}^{400} 6.20 \times 10^{4} \frac{dT}{T^3}$$

$$= 6.20 \ln\left[\ln T\right]_{300}^{400} + 1.3 \times 10^{-3} \left[T\right]_{300}^{400} - \frac{6.80 \times 10^{4}}{-2}\left[\frac{1}{T^2}\right]_{300}^{400}$$

$$= 6.20 \ln \frac{400}{300} + 1.3 \text{ Ÿ } 10^{-3} \text{ Ÿ } (400 - 300) + 3.40 \text{ Ÿ } 10^{4}$$

$$\left[\frac{1}{(400)^2} - \frac{1}{(300)^2}\right] = 1.7487 \text{ cal./degree.}$$

Problem 11:

A thermostat was maintained at 370.05 K. For nearly an hour 4.2 kJ of heat leaked through the thermostat insulation into a room where the initial temperature of air was 300.05 K. (i) What was the entropy change of the material in the thermostat ? (ii) What was the entropy change of the air in the room ? (iii) Was the process spontaneous? (Material in the thermostat is water).

Solution:

(i) Entropy change of the material in the thermostat

$$-\frac{4.2 \text{ kJ}}{370.05\text{K}} = -11.35 \times 10^{3} \text{ kj K}^{-1}$$

(ii) Entropy change of the air in the room

$$= \frac{4.2 \text{ kJ}}{300.05 \text{ K}} = 14.00 \times 10^{3} \text{ kJ K}^{-1}$$

(iii) Now, $\Delta S_{total} = \Delta S_{thermostat} + \Delta S_{room}$

$$= (-11.35 + 14.00) \text{ JK}^{-1}$$

$$= 2.65 \text{ J K}^{-1}$$

Since ΔStotal is positive, the process of leaking is spontaneous.

Problem 12:

Calculate ΔS, ΔH and ΔU for the process

$$H_2O(l, 20°C, 1 atm \rightarrow H_2O (g, 250°C, 1 atm)$$

Given the following data :

$$C_p (l) = 75.6 \; J \; K^{-1} \; mol^{-1}$$

$$C_p (g) = 36.2 \; J \; K^{-1} \; mol^{-1}$$

ΔH for vaporization of H_2O at 100°C and 1 atm is equal to 40.85 k J mol^{-1}.

Solution:

The given process may be replaced by the following processes.

(i) H_2O (l, 20°C, 1 atm) → H_2O (l, 100°C, 1 atm)

(ii) H_2O (l, 100°C, 1 atm) → H_2O (g, 100°C, 1 atm)

(iii) H_2O (g, 100°C, 1 atm) → H_2O (g, 250°C, 1 atm)

Thus we have

$$\Delta S = \Delta S_{(1)} + \Delta S_{(ii)} \Delta S_{(iii)}$$

$$= C_{p,m}(l) \ln \frac{T_2}{T_1} + \frac{\Delta H_{vap}}{T} + C_{p,m}(g) \ln \frac{T_2'}{T_1'}$$

$$= \left(75.6 \ln \frac{373}{293} + \frac{40850}{373} + 36.2 \ln \frac{523}{373}\right) JK^{-1} \; mol^{-1}$$

$$= (18.25 + 109.52 + 12.24) \; J \; K^{-1} \; mol^{-1}$$

$$= 140.01 \; J \; K^{-1} \; mol^{-1}$$

$$\Delta H = \Delta H_{(1)} + \Delta H_{(ii)} + \Delta H_{(iii)}$$

$$= C_{p,m} (l) (T_2 - T_1) + \Delta H_{vap} + C_{pm}(g) (T'_2 - T'_1)$$

$$= (75.6 \times 80 + 40850 + 36.2 \times 150) \; J \; mol^{-1}$$

$$= 52328 \;\; J \; mol^{-1}$$

$$\Delta U = \Delta U_{(1)} + \Delta U_{(ii)} \Delta U_{(iii)}$$

$$= C_{v,m} (l) (T_2 - T_1) + \Delta U_{vap} + C_{v,m}(g) (T'_2 - T'_1)$$

$$= C_{p,m}(l)(T_2 - T_1) + \{\Delta H_{vap} + \Delta_{V\,g} RT\} + \{C_{p,m}(g) - R\}(T'_2 - T'_1)$$

$$= [75.6 \times 80 + \{40850 - (8.314) (373)\}$$

$$+ (36.2 - 8.314) (150)] \; J \; mol^{-1}\}$$

$= (6040 + 37749 + 4183)$ J mol^{-1}

$= 47972$ J mol^{-1}.

Problem 13:

Describe a process through with the gas can be restored to its initial state in each of the above cases. Explain how the surroundings can be restored only in one ca. e and not in the other.

Solution:

The gas can be restored to its initial state by compressing it isothermally from 5 dm^3 to 1 dm^3. This may be done reversibly or irreversibly. When the compression is done reversibly, the surroundings in the case of (ii) will be restored whereas in case of (i), it cannot be restored whether the compressions is done reversibly or irreversibly. This is because, while on expansion, surroundings remain unchanged whereas during compression, it has to do work on the system and thus suffers changes.

Problem 14:

Derive an expression for the entropy change when a van der Waals gas is heated and expanded simultaneously.

Solution:

Taking S = f(T, V), we get

$$dF = \left(\frac{\partial S}{\partial T}\right)_V dT + \left(\frac{\partial S}{\partial V}\right)_T dV$$

$$= \frac{nC_{V,m}}{T} dT + \left(\frac{\partial P}{\partial T}\right)_V dV$$

Now for van der Waal's gas we have

$$P = \frac{n\,RT}{V - nb} - \frac{n^2 a}{V^2}$$

Therefore $\left(\frac{\partial P}{\partial T}\right)_V = \frac{n\,RT}{V - nb}$

Thus $dS = \frac{n\,C_{V,m}}{T} dT + \frac{n\,R}{V - nb} dV$

Which on integrating within the limit T_1, V_1 and T_2, V_2 gives

$$\Delta S\ nC_{V,m} \ln \frac{T_2}{T_1} + nR \ln \frac{V_2 - nb}{V_1 - nb}$$

Problem 15:

For an adiabatic process, dq = 0, If one were to write

$$\Delta S = \int \frac{dq}{T} = \int \frac{0}{T} = 0$$

Then every adiabatic process would be iso-entropic process. Comment.

Solution:

For an adiabatic process, dq – 0, and according to the first law of thermodynamics, dq = dU – dw

Thus, we will have

$$dU = dw$$

The expression of entropy change will be given as

$$\Delta S = \int \frac{dU - dw}{T}$$

$$= \int \frac{dU}{T} + \int \frac{P_{opp}}{T} dV$$

Now, it is only for reversible process the first integral is equal and opposite sign of the second integral and thus ΔS = 0. For an irreversible process, these two integrals do not have equal magnitude and thus sum is not zero. Hence ΔS ≠ 0 for an adiabatic irreversible process.

Problem 16:

A system consists of 6000 particles distributed in three energy states with equal spacing. The energy of the three states are $E_1 = 0$, $E_2 = x$ and $E_3 = 2x$. All the three slates have the same intrinsic probability g. At a certain instant there are 3000 particles in lower level, 2500 in the middle level and 500 in the upper level. Compare the relative probabilities with the distribution obtained, by the transfer of one particle from the middle to the lower level and one particle from the middle to the upper level and the original distribution.

Solution:

Let P_1 and P_2 be the probabilities in the two cases

In the first case,

$$N = 6000,$$

$$n_1 = 3000,$$

$$n_2 = 2500,$$

$$n_3 = 500.$$

$$P_1 = \frac{g^N}{n_1!n_2!n_3!}$$

$$P_1 = \frac{g^{6000}}{3000!2500!500!} \quad ...(i)$$

In the second case

$$N = 6000,$$

$$n_1 = 3001,$$

$$n_2 = 2498,$$

$$n_3 = 501.$$

$$P_1 = \frac{g^N}{n_1!n_2!n_3!}$$

$$P_1 = \frac{g^{6000}}{3001!2498!501!}$$

Dividing (ii) by (i)

$$\frac{P_2}{P_1} = \frac{3000!2500!500!}{3001!2498!501!}$$

$$\frac{P_2}{P_1} = \frac{2500 \times 2499!}{3001 \times 501!}$$

$$\frac{P_2}{P_1} = 4.157$$

$$\approx 4.2.$$

It means the transfer of one particle from the middle to the upper and one particle from the middle to the lower state has changed the probability by a factor 4.2. This shows that both these distributions are not near the equilibrium state.

Problem 17:

A thermostat was maintained at 370.05 K. For nearly an hour 4.2 kJ of heat leaked through the thermostat insulation into a room where the initial temperature of air was 300.05 K.

(i) What was the entropy change of the material in the thermostat?

(ii) What was the entropy change of the air in the room ?

(iii) Was the process spontaneous? (Material in the thermostat is water).

Solution:

(i) Entropy change of the material in the thermostat

$$-\frac{4.2\text{ kJ}}{370.05\text{K}} = -11.35\times 10^3\text{ kj K}^{-1}$$

(ii) Entropy change of the air in the room

$$= \frac{4.2\text{ kJ}}{300.05\text{ K}} = 14.00 \times 10^3\text{ kJ K}^{-1}$$

(iii) Now

$$\Delta S_{total} = \Delta S_{thermostat} + \Delta S_{room}$$
$$= (-11.35 + 14.00)\text{ JK}^{-1}$$
$$= 2.65\text{ J K}^{-1}$$

Since ΔStotal is positive, the process of leaking is spontaneous.

Problem 18:

At low temperature, a nonideal gas follows the relation PV = RT – a/V where a = 0.3636 Nm⁴ mole⁻². Calculate the work done by one mole of this gas in expanding form 0.224 litre to 22.4 litre at 400 K. Compare this result with the work done in the corresponding expansion of an ideal gas.

Solution:

For an ideal gas, reversible work done

$$= 2.303\text{ n RT log}\frac{V_2}{V_1}$$

$$V_1 = 0.224\text{ litres mole}^{-1} = 22.4 \times 10^{-5}\text{ m}^3\text{ mole}^{-1}$$

$$P_2 = 22.4\text{ litres mole}^{-1} = 22.4 \times 10^{-3}\text{ m}^3\text{ mole}^{-1}$$

$$R = 8.314\text{ J K}^{-1}\text{ mole}^{-1}$$

$$W_{rew} = (2.303)\ (1\text{ mole})\ (8.314\text{ J K}^{-1}\text{ mole}^{-1})\ (400\text{ K})$$
$$\times \log\frac{22.4\times 10^{-3}}{22.4\times 10^{-5}}$$

$= (2.303)\ (1\ \text{mole})\ (8.314\ \text{J K}^{-1}\ \text{mole}^{-1})\ (400\ \text{K})\ \log 100$
$= 15320\ \text{J mole}^{-1}$
$= 15.32\ \text{kJ mole}^{-1}.$

For the non-ideal gas PV = RT – a/V

$$\text{Therefore, the work done} = \int_{V_1}^{V_2} P\,dV = \int_{V_1}^{V_2} \left\{\frac{RT}{V} - \frac{a}{V_2}\right\} dV$$

$$= 2.303\ RT \log \frac{V_2}{V_1} + a\left\{\frac{1}{V_2} - \frac{1}{V_1}\right\}$$

The value of first part is from derived eq. (1) as 15.32 kJ mole^{-1} and of another part can be derived as

$$0.3636\ \text{Nm}^4\ \text{mole}^{-2}\left\{\frac{1}{22.4 \times 10^{-3}\ \text{m}^3\ \text{mole}^{-1}} - \frac{1}{22.4 \times 10^{-5}\ \text{m}^3\ \text{mole}^{-1}}\right\}$$

$= 1.604\ \text{kJ mole}^{-1}.$

Thus work done $= 15.320\ \text{k Joule}^{-1} - 1.604\ \text{k Joule}^{-1}$
$= 13.716\ \text{kJ mole}^{-1}.$

Problem 19:

What will be the work done when 65.38 g of zinc dissolves in hydrochloric acid in case of an open beaker and closed beaker at 300K.

Solution:

The process of dissolution of zinc in hydrochloric acid can be given as

$$Zn(s) + 2HCI\ (aq) = ZnCl_2(aq) + H_2(g)$$

From this equation it is clear that for each gram atom of zinc dissolved we will obtain 1 mole of Hydrogen gas. As a result of liberated gas, on the surrounding atmosphere the work performed will be equal to PΔV. If the initial volume of the system is neglected in comparison to the total volume of the gas produced and the gas shows ideal behaviour, then the work done in the open beaker is

$W = P\Delta V \approx PV_{H2}$
$= nH2\ RT\ [\because\ T = 300\ k\ R = 8.314\ \text{Jk}^{-1}\ \text{mole}^{-1}]$
$= (1\ \text{mole})\ (8.314\ \text{JK}^{-1}\ \text{mol}^{-1})\ (300\ \text{K})$

$= 2.4942$ kJ.

If the reaction takes place in a closed beaker there is change in the volume *i.e.*, $\Delta V = 0$ hence work done is zero.

Problem 20:

Calculate the work done when one mole of sulphur dioxide gas expands isothermally and reversibly at 300 k from 2.46×10^{-3} m^3 to 24.6×10^{-3} m^3. Assuming that the gas obeys

(a) ideal gas equation and

(b) van der Waals equation

$$\left\{ \begin{array}{l} a = 0.6799 \text{ nm}^4 \text{ mole}^{-2} \\ b = 0.564 \times 10^{-3} \text{ mole}^{-1} \end{array} \right\}$$

Solution:

(a) The reversible work done by 1 mole of gas when it expands form

$$2.64 \times 10^{-3}\text{m}^3 \text{ to } 24.6 \times 10^{-3}\text{m}^3 = 2.303 \text{ RT} \log \frac{V_2}{V_1}$$

$$= (2.303)\ (1 \text{ mole})\ (8.314 \text{ J K}^{-1} \text{ mol}^{-1})\ (300 \text{ K}) \times \log 10$$

$$\left\{ \because \log \frac{V_2}{V_1} = \log 10 \right\}$$

$$= (2.303)\ (8.314)\ (300 \text{ K})\ \log 10 \text{ J mol}^{-1}$$

$$= 5744 \text{ J mol}^{-1}$$

(b) When the gas obeys van der Waal's equation, work done by 1 mole of gas is

$$= 2.303 \text{ RT} \log \frac{V_2 - b}{V_1 - b} + a \left\{ \frac{1}{V_2} - \frac{1}{V_1} \right\}$$

After substituting the values of various quantities and solving, the work done is

$$= 5794 \text{ J mol}^{-1} - 252.4 \text{ J mol}^{-1}$$

$$= 5541.6 \text{ J mol}^{-1}$$

$$= 5.542 \text{ kJ mol}^{-1}$$

Problem 21:

One mole of a van der Walls's gas is allowed to expand isothermally and reversibly from a volume of 1 litre to 50 litre at 0°C. Calculate w, *q, Δ E and Δ H. van der Waal's constants are a = 6.5atm l^2mole $^{-2}$ = 0.056 litre mole $^{-1}$ and R = 0.082 l atom deg^{-1} mole^{-1}.*

Solution:

Here $V_1 = 1$ litre,

$V_2 = 50$ litres, n = 1 mole

T = 273 + 0 = 273°k,

R = 0.082 l-atm deg^{-1} mole^{-1}

a = 6.5 atm l^2 mole^{-2},

b = 0.056 litre mole^{-1}

$$w = 2.303 \text{ nRT} \log\left(\frac{V_2 - nb}{V_1 - nb}\right) + an^2\left(\frac{1}{V_2} - \frac{1}{V_1}\right)$$

$$= 2.303 \times 1 \times 0.082 + 273 \,[\log(50 - 0.056) - \log(1 - 0.056) + 6.5\left(\frac{1}{50} - \frac{1}{1}\right)$$

= 82.47 l atm.

$$\Delta E = -an^2\left(\frac{1}{V_2} - \frac{1}{V_1}\right) = 6.5 \times \frac{49}{50}$$

$$q = 2.303\text{nRT} \log \frac{V_2 - b}{V_1 - b}$$

= ΔE + w = 884 l atm.

Problem 22:

The van der Waal's constant a and b for hydrogen in litre atmosphere units are 0.246 and 2.67 × 10^{-2} respectively. Calculate the inversion temperature of hydrogen.

Solution:

we know $T_i = 2a/Rb$

or
$$T_i = \frac{2 \times 0.246}{0.0821 \times 0.0267}$$

= 224.5°k = (224.5 – 273)

= – 48.5°C.

Problem 23:

How much heat is required to raise the temperature of 1 mole oxygen from 300K to 1300 K at constant pressure.

$$C_p = 6.095 + 3.325 \times 10^{3} T - 1.017 \times 10^{6} T^2.$$

Solution:

$$H_2 - H_1 \int_{T_t}^{T_2} C_p \, dT$$

$$T_1 = 300 \text{ k}, T_2 = 100 \text{ k}$$

$$\therefore \quad H_{1300} - H_{300}$$

$$= \int_{300}^{1300} \left(6.095 + 3.253 + 10^{-3} \, T - 1.017 \times 10^{-6} \, T^2\right)$$

[Substituting the value of C_p]

$$\because \quad H_{1300} - H_{300} = 6.095\ (1300 - 300) + \frac{3.253 \times 10^{-3}}{2}$$

$$\times [(1300)^2 - (300)^2] - 1/3 \times 1.017 \times 10^{-6} [(1000)^2 - (300)^2]$$

$$= 6095 + 2602.4 - 735.63$$

$$= 7961.77 \text{ cal/mole.}$$

Problem 24:

For equation (P _ a/V²)˙ V = RT, prove that (i) dP is an exact differential, (ii) P is a state function and

$$\text{(iii)} \left(\frac{\partial P}{\partial T}\right)_V \left(\frac{\partial T}{\partial V}\right)_P \left(\frac{\partial V}{\partial T}\right)_T + 1 = 0.$$

Solution:

(i) For dP to be an exact differential, it is be proved that

$$\frac{\partial^2 P}{\partial V \partial T} = \frac{\partial^2 P}{\partial T \partial V} \qquad \text{...(1)}$$

$$(P + a/V^2)\ V = RT$$

or $$P = \frac{RT}{V} - \frac{a}{V^2} \qquad \text{...(2)}$$

When the above equation is differentiated with respect to V at constant temperature T, we obtain

$$\left(\frac{\partial P}{\partial V}\right)_T = -\frac{RT}{V^2} + \frac{2a}{V^3}$$

When the above equation is differentiated with respect to T at constant V, we obtain

$$\frac{\partial^2 P}{\partial T \partial V} = -\frac{RT}{V^2} \qquad ...(3)$$

When equation (2) is differentiated with respect to T at constant volume V, we obtain

$$\left(\frac{\partial P}{\partial T}\right)_V = \frac{R}{V}$$

When the above equation is further differentiated with respect to V at constant T, we obtain

$$\frac{\partial^2 P}{\partial T \partial V} = -\frac{R}{V^2} \qquad ...(4)$$

From equations (3) and (4), we get

$$\frac{\partial^2 P}{\partial T\, \partial V} = \frac{\partial^2 P}{\partial T\, \partial V} = -\frac{R}{V^2} \qquad ...(5)$$

Thus, we have proved condition, (1). This shows that dP is an exact differential.

(ii) In the said question, first of all a change occurs in volume at constant temperature followed by a change in temperature at constant volume, thereby giving the final equation (3).

Again, a change in temperature is considered at constant volume, followed by a change in volume at constant temperature, thereby giving the final equation (4). Both equations (3) and (4) are same, thereby revealing that the final change in P remains the same irrespective of the difference in intermediate changes. Therefore, P is a state function as changes in P have been found to depend upon the intial and final states of the system.

(iii) Equation (2) is as follows :

$$P = \frac{RT}{V} - \frac{a}{V^2}$$

When the above equation is differentiated, we obtain :

$$dP = \frac{R}{V}\, dT - \frac{RT}{V^2}\, dV + \frac{2a}{V^3}\, dV$$

$$= \frac{R}{V}\, dT - \left(\frac{RT}{V^2} - \frac{2a}{V^3}\right) dV \qquad ...(6)$$

If volume is constant, dV = 0 : equation (6) becomes as follows :

$$dP = \frac{R}{V}\, dT \text{or} \left(\frac{\partial P}{\partial V}\right)_V = \frac{R}{V} \quad ...(7)$$

If pressure is constant, dP = 0; equation (6) becomes as follow :

$$\frac{R}{V}\, dT - \left(\frac{RT}{V^2} - \frac{2a}{V^3}\right) dV = 0$$

or
$$\left(\frac{\partial T}{\partial V}\right)_P = \left(\frac{RT}{V^2} - \frac{2a}{V^3}\right) \Big/ \frac{R}{V} \quad ...(8)$$

If temperature is constant, dT = 0; equation (6) becomes as follows:

$$\left(\frac{\partial T}{\partial V}\right)_T = -\frac{1}{\left(\frac{RT}{V^2} - \frac{2a}{V^3}\right)} \quad ...(9)$$

On multiplying equations (7), (8) and (9), we obtain

$$\left(\frac{\partial p}{\partial T}\right)_V \left(\frac{\partial T}{\partial V}\right)_P \left(\frac{\partial V}{\partial P}\right)_T = -\frac{R}{V} \times \frac{\left(\frac{RT}{V^2} - \frac{2a}{V^3}\right)}{\left(\frac{RT}{V^2} - \frac{2a}{V^3}\right)} \times \frac{1}{R} = -1$$

or
$$\left(\frac{\partial p}{\partial T}\right)_V \left(\frac{\partial T}{\partial V}\right)_P \left(\frac{\partial V}{\partial P}\right)_T + 1 = 0$$

This is cyclic rule.

Problem 25:

For an ideal gas (PV = nRT) show that (1/T) is an integrating factor for dw = P dV.

Solution:

$$dw = P\, dV \quad ...(1)$$

$\because$
$$PV = nRT$$

$\therefore$
$$p\, dV + V\, dP = nR\, dT$$

and
$$dV = \frac{nR}{P} dT - \frac{V}{P} dP \quad ...(2)$$

On substituting the expression for dV in equation (1), we obtain

$$dw = nR\, dT - V\, dP$$

$$z = nR\,dT - \frac{nRT}{P}dP$$

Now $\frac{\partial}{\partial P}(nR)_T = 0$

and $\frac{\partial}{\partial T}\left(-\frac{nRT}{P}\right)_P = -\frac{nR}{P}$

Hence, dw is inexact differential.

Suppose f = 1/T be an integrating factor. Then

$$f\,.\,dw = \frac{1}{T}nR\,dT - \frac{nR}{P}dP$$

$$\therefore \qquad \frac{\partial}{\partial P}\left[\frac{nR}{T}\right]_T = 0$$

and $\frac{\partial}{\partial T}\left[\frac{nR}{P}\right]_T = 0$

Thus f. dw$\frac{nR}{T}dT - \frac{nR}{P}dP$ is an exact differential while 1/T is an integrating factor.

Problem 26:

The molar heat capacity at capacity at constant pressure of solid magnesium (0° – 600°C) is expressed by

CP = 6.20 + 1.30 × 10 – 3 T – 6.80 × 104 T – 2

Calculate the increase in entropy when 1 gm atom of metal is heated from 27°C to 127°C at constant pressure.

Solution:

At constant pressure,

$$\Delta S = \int_{T_1}^{T_2} \frac{C_P}{T}\,dT$$

$$\therefore \quad \Delta S = \int_{300}^{400} \frac{\left(6.20 + 1.30 \times 10^{-3}\,T - 6.80 \times 10^4\,T^{-2}\right)dT}{T}$$

$$\int_{300}^{400} 6.20\frac{dT}{T} + \int_{300}^{400} 1.3 \times 10^{-3} dT - \int_{300}^{400} 6.20 \times 10^4 \frac{dT}{T^3}$$

$$= 6.20\ \ln\left[\ln T\right]_{300}^{400} + 1.3 \times 10^{-3}\left[T\right]_{300}^{400} - \frac{6.80 \times 10^4}{-2}\left[\frac{1}{T^2}\right]_{300}^{400}$$

$$= 6.20 \ln \frac{400}{300} + 1.3 \times 10^{-3} \times (400 - 300) + 3.40 \times 10^4$$

$$\left[\frac{1}{(400)^2} - \frac{1}{(300)^2}\right]$$

$$= 1.7487 \text{ cal./degree.}$$

Problem 27:

Calculate the increase in entropy of three moles of hydrogen as it changes from 300°K atm to 1000°K and 1 atm (C_P = 7 cal./degree/ mole).

Solution:

Entropy change ΔS for n moles

$$= n\ C_P \ln \frac{T_2}{T_1}\ n\ R \ln \frac{P_2}{P_1}$$

$$= 3 \times 7 \times 2.303 \log \frac{1000}{300} - 3 \times 2 \times 2.303 \log \frac{10}{0.1}$$

$$= 25.287996 - 13.818$$

$$= 11.469996 \text{ cal./degree.}$$

Problem 28:

Calculate ΔS, ΔH and ΔU for the process

$$H_2O(l, 20°C, 1\ atm \rightarrow H_2O\ (g, 250°C, 1\ atm)$$

Given the following data :

$$C_P\ (l) = 75.6\ J\ K^{-1}\ mol^{-1}$$

$$C_P\ (g) = 36.2\ J\ K^{-1}\ mol^{-1}$$

ΔH for vaporization of H_2O at 100°C and 1 atm is equal to 40.85 k J mol^{-1}.

Solution:

The given process may be replaced by the following processes.

(i) H_2O (l, 20°C, 1 atm) → H_2O (l, 100°C, 1 atm)

(ii) H_2O (l, 100°C, 1 atm) → H_2O (g, 100°C, 1 atm)

(iii) H_2O (g, 100°C, 1 atm) → H_2O (g, 250°C, 1 atm)

Thus we have

$$\Delta S = \Delta S_{(1)} + \Delta S_{(ii)} \Delta S_{(iii)}$$

$$= C_{p,m}(l) \ln \frac{T_2}{T_1} + \frac{\Delta H_{vap}}{T} + C_{p,m}(g) \ln \frac{T_2'}{T_1'}$$

$$= \left(75.6 \ln \frac{373}{293} + \frac{40850}{373} + 36.2 \ln \frac{523}{373}\right) JK^{-1} \ mol^{-1}$$

$$= (18.25 + 109.52 + 12.24) \ J \ K^{-1} \ mol^{-1}$$

$$= 140.01 \ J \ K^{-1} \ mol^{-1}$$

$$\Delta H = \Delta H_{(1)} + \Delta H_{(ii)} + \Delta H_{(iii)}$$

$$= C_{p,m} (l) (T_2 - T_1) + \Delta H_{vap} + C_{pm}(g) (T'_2 - T'_1)$$

$$= (75.6 \times 80 + 40850 + 36.2 \times 150) \ J \ mol^{-1}$$

$$= 52328 \ \ J \ mol^{-1}$$

$$\Delta U = \Delta U_{(1)} + \Delta U_{(ii)} \Delta U_{(iii)}$$

$$= C_{v,m} (l) (T_2 - T_1) + \Delta U_{vap} + C_{v,m}(g) (T'_2 - T'_1)$$

$$= C_{p,m}(l)(T_2 - T_1) + \{\Delta H_{vap} + \Delta_{v\,g}RT\} + \{C_{p,m}(g) - R\}(T'_2 - T'_1)$$

$$= [75.6 \times 80 + \{40850 - (8.314)(373)\}$$

$$+ (36.2 - 8.314)(150)] \ J \ mol^{-1}\}$$

$$= (6040 + 37749 + 4183) \ J \ mol^{-1}$$

$$= 47972 \ J \ mol^{-1}.$$

Problem 29:

For N = 400, find the ratios of the probabilities for (300, 100) and (200, 200) distribution in two identical boxes.

Solution:

We have

$$P(300, 100) = \frac{400!}{300! \, 100!} \frac{1}{2^{400}}$$

$$P(200, 200) = \frac{400!}{200! \, 200!} \frac{1}{2^{400}}$$

$$\frac{P(300, 100)}{P(200, 200)} = \frac{200! \, 200}{300! \, 100!} \cong 10^{-5}.$$

Problem 30:

Eight similar coins are tossed, find the probability of three heads and t tails.

Solution:

We know $$P(3, 5) = \frac{^8C_3}{2^8} = \frac{8\times7\times6}{3\times2\times1\times2^8}$$

or $$P(3, 5) = \frac{7}{32}$$

Problem 31:

Two bodies at different temperatures are connected by a wire with infinitesimal thermal conductance so that heat flows infinitely slowly from the hot bodies to the cold body until they are at the same temperature. Is this process reversible? Explain.

Solution:

In a reversible process, at any stage, the external condition responsible for the process to occur differs from the internal condition by an infinitesimal amount. The given transfer of heat is not a reversible process as the two bodies are at different temperatures. Moreover, the reversible processes are characterized by the fact that when the system is restored to its original state by traversing the forward sequence of steps in the reverse order, then not only the system but also the surroundings are restored to their original states. In the given process, heat flow from a hot body to a cold body does not disturb the surroundings but if we wish to carry out the reverse process, the surroundings will have to do some work. Hence, the surroundings are not brought back to the original state in the reverse process. Thus, the given process is not a reversible one.

Problem 32:

How would the energy of an ideal gas change if it is made to expand into vacuum at constant temperature ?

Solution:

For the expansion against vacuum $P_{opp} = 0$

Hence $$w = -P_{opp}\,\Delta V = 0$$

Since the expansion is at constant temperature no heat is supplied to the system, *i.e.*, q = 0. Now, from the first law of thermodynamics, we can write

$$\Delta E = q + w = 0$$

that is, the energy of an ideal gas does not change if it is made to expand into vacuum at constant temperature. This also follows from the fact that the energy of an ideal gas depends only on temperature and since temperature remains constant, its energy also remains constant.

Problem 33:

Calculate the change of entropy for the process

$$H_2O\ (l,\ 373\ K) \rightarrow H_2O\ (v,\ 373\ K)$$

Here $\Delta H_{vap} = 40850\ J\ mol^{-1}$ *at 373 K.*

Solution:

$$\Delta S_{vap} = \frac{\Delta H_{vap}}{T_b}$$

$$= \frac{40850\ J\ mol^{-1}}{373\ K}$$

$$= 109.51742\ JK^{-1}\ mol^{-1}$$

But $\Delta S_{vap} = S\ (vapour) - S\ (liquid)$

$$= 109.5\ JK^{-1}\ mol^{-1}$$

$\therefore$ $S\ (vapour) = S\ (liquid) + 109.5\ JK^{-1}\ mol^{-1}$

Problem 34:

Calculate the work down when 50 g of iron dissolves in hydrochloric acid at 25oC in :

(i) A closed vessel, and

(ii) An open beaker.

(iii) In which case would maximum amount of work be done ?

Solution:

$$\text{Amount of iron} = \frac{50\ g}{55.85\ g\ mol^{-1}} = \frac{50}{55.85}\ mol$$

(i) In a closed vessel, $\Delta V = 0$. Hence,

$$dw = -PdV = 0$$

that is, no work is involved when iron dissolves in hydrochloric acid in a closed vessel.

(ii) The reaction between iron and hydrochloric acid is

$$Fe + 2HCl \rightarrow FeCl_2 + H_2(g)$$

Thus, the liberated hydrogen gas pushes the atmospheric air and thus involves the work of expansion. In this case, we will have

$$w = -P_{ext}\, \Delta V$$
$$= -P_{ext}\, V_{H2}$$

where V_{H2} is the volume of hydrogen released. According to the ideal gas law, we will have

$$w = -P_{ext}\,(n_{H2}RT/P_{ext})$$
$$= -n_{H2}RT$$
$$= -(50/55.85)\text{ mol }(8.314\text{ J K}^{-1}\text{ mol}^{-1})(298\text{ K})$$
$$= -2.2\text{ kJ}.$$

Problem 35:

The opposite faces of a metal plate of 0.2 cm thickness an at a difference of temperature of 109°C and the area of the plate is 200 sq cm. Find the quantity of heat that will flow through the plate in one minute if K = 0.2 CGS units.

Solution:

Here,

$$K = 0.2$$
$$A = 200\text{ sq cm}$$
$$d = 0.2\text{ cm}$$
$$(\theta_1 - \theta_2) = 100°C$$
$$t = 60s$$
$$Q = \frac{KA(\theta_1 - \theta_2)t}{d}$$
$$= \frac{0.2 \times 200 \times 100 \times 60}{0.2}$$
$$= 12 \times 10^5\text{ cal.}$$

Problem 36:

A bar of length 30 cm and uniform area of cross-section 5 cm^2 consists of two halves AB of copper and BC of iron welded together at B. The end A is maintained at 200°C and the end C at 0°C. The sides of the bar are thermally insulated. Find the rate of flow of heat along the bar when the steady state is reached. Thermal conductivity of copper if 0.9 and thermal conductivity of iron is 0.12 CGS units.

Solution:

Suppose the temperature at the interface. B is θ after the steady state is reached.

$$\therefore \quad \frac{K_1A(200-\theta)}{d} = \frac{K_2A(\theta-0)}{d}$$

$$0.9(200 - \theta) = 0.12\ \theta$$

$$\theta = 176.5°C.$$

After the steady state is reached, the rate of flow of heat is the same in both the bars.

$$\therefore \quad Q = \frac{K_1A(\theta-0)}{d}$$

$$Q = \frac{0.9\times5\times(200-176.5)}{15}$$

$$Q = 7.05 \text{ cal/s.}$$

Problem 37:

Find the time in which a layer of ice 3 cm thick on the surface of a pond will increase its thickness by 1 mm when the temperature of the surrounding air is –20°C.

Thermal conductivity of ice = 0.005

Latent heat of ice = 80 cal/g

Density of ice at 0°C = 0.91 g/cm^3.

Solution:

$$\int dt = \frac{\rho L}{K\theta}\int x dx$$

$$t = \frac{\rho L}{2K\theta}[x_2^2 - x_1^2]$$

Here, $\rho = 0.91$ g/cm^3, L = 80 cal/g

$$K = 0.005,\ \theta = 20^{\circ}C$$

$$x_1 = 3\text{ cm},\ x_2 = 3.1\text{ cm}$$

$$t = \frac{0.91\times80}{2\times0.005\times20}[(3.1)^2-(3)^2]$$

$$= 222.04\text{ s},\ = 3\text{ min } 42\text{ s approximately.}$$

Problem 38:

An ice box is built of wood 1.76 cm thick, lined inside with cork 3 cm thick. If the temperature of the inner surface of the cork is 0°C and that of the outer surface of wood is 12°C, what is the temperature of the interface ? The thermal conductivity of wood and cork are 0.0006 and 0.00012 CGS units respectively.

Solution:

Suppose the temperature of the interface is 9 after the steady state is reached

$$\therefore \quad \frac{K_1A(12-\theta)}{d_1} = \frac{K_2A(\theta-0)}{d_2}$$

Here

$$K_1 = 0.0006,$$
$$K_2 = 0.00012$$
$$d_1 = 1.75\text{ cm},$$
$$d_2 = 3\text{ cm}$$

$$\frac{0.0006(12-\theta)}{1.75} = \frac{0.00012(\theta)}{3}$$

$$\theta = 10.74^{\circ}C.$$

Problem 39:

Since CV = (∂q/∂T) V by definition, one often writes without restriction dV = C_VdT. This is generally not true. Explain why.

Solution:

If we take U = f)V, T), then we will have

$$dU = \left(\frac{\partial U}{\partial V}\right)_T dV + \left(\frac{\partial U}{\partial T}\right)_V dT$$

$$= \left(\frac{\partial U}{\partial V}\right)_T . dV + C_V\, dT$$

Thus, if one writes $dV = C_V\, dT$ without restriction then this statement is not complete. This is applicable for a system in which $(\partial U/\partial V)_T = 0$. An ideal gas is such an example. Alternatively, the system may be undergoing change at constant volume, for which, $(\partial U/\partial V)_T\, dV$ is zero and thus $dV = C_V\, dT$.

Problem 40:

Is the equation PV^{γ} = constant valid for all processes ? Explain your answer.

Solution:

The expression PV^{γ} = constant is valid only for adiabatic reversible volume change (*i.e.*, expansion or compression) involving an ideal gas. For adiabatic process, dq = 0.

Hence

$$dU = dw$$

$$C_V dT = - P_{ext}\, dV$$

For revisable process, $P_{ext} = P_{int}$. Now replacing P_{int} = nRT/V and separating the variables followed by integration, we get

$$\frac{Cv}{n} \int \frac{dT}{T} = - R \int \frac{dV}{V}$$

or $$C_{V.m} \ln \frac{T_2}{T_1} = - R \ln \frac{V_2}{V_1}$$

or $$\left(\frac{T_2}{T_1}\right)^{C_{v,m}} = \left(\frac{V_2}{V_1}\right)^{-R}$$

Replacing T_2/T_1 by $P_2 V_2/P_1 V_1$ and rearranging, the expression would yield

$$P_2 V^{\gamma}_2 = P_1 V^{\gamma}_1$$

that is $$PV^{\gamma} = \text{constant.}$$

Problem 41:

Under what conditions does the adiabatic expansion of gas become a constant energy process ?

Solution:

If the adiabatic expansions done against a vacuum then in this process energy

On comparing the coefficients in Eqs (1) and (2), we have

$$\left(\frac{\partial E}{\partial T}\right)_V = T \qquad ...(1)$$

and $$\left(\frac{\partial E}{\partial V}\right)_S = -P \qquad ...(2)$$

Eq (2) can be written as

$$dE = T\,dS \qquad ...(3)$$

For a given temperature, the above equation on integration becomes as follows :

$$\int dE = \int TdS + \text{Integration constant}$$

$$E = TS + A \qquad ...(4)$$

or $$E = A - TS$$

where A is an integration constant. It is also known as Helmholtz free energy function.

(d) Enthalpy as a Function of S and P

As H is a function of S and P, we can write

$$H = f(S, P)$$

or $$dH = \left(\frac{\partial H}{\partial S}\right)_P dS + \left(\frac{\partial H}{\partial P}\right)_S dP \qquad ...(5)$$

We also know, $$dH = T\,dS + V\,dP \qquad ...(6)$$

On comparing the coefficients in Eqs (5) and (6), we get

$$\left(\frac{\partial H}{\partial S}\right)_P = T \qquad ...(7)$$

$$\left(\frac{\partial H}{\partial P}\right)_S = V \qquad ...(8)$$

When Eq. (7) is integrated, we get a new function

$$\int dH = \int TdS + \text{constant}$$

$$= \int TdS + G$$

or $$H = TS + G$$

or $$G = H - TS$$

where G is an integration constant and is the same as the Gibbs free energy function.

Problem 42:

Two moles of an ideal gas at 300 K expand reversibly and isothermally from 4×10^{-2} m^3 to 8×10^{-2} m^3. Calculate the entropy change for the gas. How would you account for the fact that ΔS is not zero although the process is reversible ?

Solution:

If the process is reversible isothermal it means that $\Delta H = 0$, hence the entropy change is given as follows :

$$\Delta S_{sys} = \frac{Q_{rev}}{T} = \frac{W_{rev}}{T} = nR \ln \frac{V_2}{V_1}$$

$$= (2 \text{ mol})(8.314 \text{ JK}^{-1} \text{ mol}^{-1}) \left(2.303 \log \frac{8 \times 10^{-2}}{4 \times 10^{-2}} \right)$$

$$= 2 \times 8.314 \times 2.303 \log 2 \text{ JK}^{-1}$$

$$= 2 \times 8.314 \times 2.303 \times 0.3010 \text{ JK}^{-1}$$

$$= 11.526579 \text{ JK}^{-1}.$$

The total entropy change (system and its surroundings) will be zero for a reversible process.

Problem 43:

Calculate the entropy changes involved in the conversion of 1 mole of ice at 0°C and 1 atm to liquid at 0°C and 1 atm [The enthalpy of fusion per mole of ice is 6008 J mol⁻¹].

Solution:

$$H_2O(S) \rightarrow H_2O(l)$$

$$\Delta S_{fus} = \frac{\Delta H_{fus}}{T_m}$$

$$= \frac{6008 \text{ J mol}^{-1}}{273 \text{ K}} = 22.007326 \text{ JK}^{-1}.$$

Problem 44:

Increase in volume of a gas from a given decrease of pressure is more in an isothermal expansion than in an adiabatic expansion. Explain.

Solution:

In an isothermal expansion, temperature remains the same whereas in adiabatic expansion temperature decreases. Since the final pressure is them same, it follows from Charles law ($V \propto T$, p constant) that

$$V_{iso} > V_{adi}$$

Since

$$\Delta V_{iso} = V_{iso} - V_{intial}$$

$$\Delta V_{adi} = V_{adi} - V_{intial}$$

it follows that.

Problem 45:

Evaluate the entropy change for the following reversible process.

$$1 \text{ mole Sn } (\propto, 13^oC \rightleftharpoons 1 \text{ mole Sn } (b, 13^oC)$$

where $\Delta H_{trans} = 2090 \text{ J mol}^{-1}$

Solution

$$T_{trans} = 273 + 13 = 286 \text{ K}$$

$$\Delta S_{trans} = \frac{\Delta H_{trans}}{T_{trans}}$$

$$= \frac{2090 \text{ J mol}^{-1}}{286 \text{ K}}$$

$$= 7.3076923 \text{ JK}^{-1} \text{ mol}^{-1}$$

Problem 46:

Calculate the entropy change when two moles of; an ideal gas expands reversibly from an initial volume of 2 dm³ to a total volume of 20 dm³ at a constant temperature of 298 K.

Solution:

We know, $\Delta S = nR \ln\left(\frac{V_2}{V_1}\right)$

$$= 2.303 \text{ nR} \log\left(\frac{V_2}{V_1}\right)$$

$$= 2.303 \times 2 \times 8.314 \text{ J JK}^{-1} \text{ mol}^{-1} \log\left(\frac{20 dm^3}{2 dm^3}\right)$$

$$= 2.303 \times 2 \times 8.314 \times \log \text{ JK}^{-1} \text{ mol}^{-1}$$

$$= 38.294284 \text{ JK}^{-1} \text{ mol}^{-1}.$$

Problem 47:

The mixing of gases is always accompanied by an increase in entropy. Show that in the formation of a binary mixture of two ideal gases the maximum entropy increase results when $X_1 = X_2 = 0.5$.

Solution:

For a binary ;mixture, the entropy per mole of the mixture formed is given by

$$\Delta S \text{ mixing} = -R[X_1 \ln X_1 + X_2 \ln X_2]$$
$$= -R[X_1 \ln X_1 + (1 - X_1) \ln (1 - X_1)]$$

For entropy of mixing to be maximum, the first derivative, $\frac{\delta(\Delta S \text{ mixing})}{\delta X_1}$ should be zero and the second derivative should be negative. Differentiating, ΔS mixing with respect to X_1 and equating it to zero yields

$$\frac{\delta(\Delta S \text{ mixing})}{\delta X_1} = -R\left[\ln X_1 + \frac{X_1}{X_1} + \frac{1 - X_1}{1 - X_1}(-1) + (-1)\ln(1 - X_1)\right]$$

$$= 0$$

$$\ln X_1 + 1 - 1 - \ln(1 - X_1) = 0$$

$$\ln \frac{X_1}{1 - X_1} = 0$$

$$X_1 = 1 - X_1$$

$$X_1 = 1/2.$$

Problem 48:

Under what conditions does the isothermal expansion of a gas become a free expansion process?

Solution:

The isothermal expansion of a gas becomes a free expansion process only when it expands against a vacuum, *i.e.*, $p_{ext} = 0$. Moreover, the gas must be an ideal gas in which case $\Delta E = 0$ at constant temperature.

Problem 49:

Is every adiabatic system isolated? In what ways can the energy of an adiabatic system change?

Solution:

An isolated system is the one in which no interaction with its surroundings takes place. Neither energy nor matter can be transferred to or from it. An adiabatic system involves no exchange of heat from its surroundings, *i.e.* for such a system dq = 0 and, according to the first law of thermodynamics, dE =dw. Thus, the energy of an adiabatic system can be changed by either doing work on the system or by getting work done by the system.

Problem 50:

One mole of a gas is expanded isothermally and reversibly at 27° C form a volume of 5 liters. Calculate q, w, ΔE and ΔH.

Solution:

We know

$PV = nRT, \qquad n = 1$

State : Initial

$T_1 = 27 + 273.15 = 300.15$ K

$V_1 = 5$ litres

Final

$T_2 = T_1 = 300.15$K

$V_2 = 10$ litres

(1) Calculation of w :

$$w = \int_{V_1}^{V_2} P\,dV = \int_{V_1}^{V_2} \frac{RT}{V} dV = RT \ln \frac{V_2}{V_1}$$

$$= 2.303 \times 8.314 \times 300 \log (10/5)$$

$$= 2.303 \times 8.314 \times 300 \times \log^2$$

$$= 2.303 \times 8.314 \times 300 \times 0.3010 \quad [\log 2 = 0.3010]$$

$$w = 1729 \text{ J mol}^{-1}$$

(2) Calculation of ΔE :

$$\Delta E = E_2 - E_1$$

For isothermal-reversible expansion of an ideal gas $E_2 = E_1$ and therefore

$$\Delta E = 0.$$

(3) Calculation of q:

$$q = \Delta E + w \text{ (first law)}$$

$$= 0 + w$$

$$q = w = 1729 \text{ J}$$

(4) Calculation of ΔH

$$\Delta H = \Delta E + \Delta (PV) = \Delta E + \Delta(RT)$$

$$0 + 0 = 0$$

Since $\Delta E = 0$ and $\Delta(RT) = 0$; R and T being constant

Problem 51:

Calculate the minimum work which must be done to compress 1/2 mole of oxygen at 300 k from a pressure of 2 atm to a pressure of 200 atm.

Solution:

As it is required to calculate the minimum work of compression, the change, is reversible, *i.e.*, the gas gets compressed by gradual increase of pressure so that the external pressure becomes only infinitesimally greater than the pressure of the gas, at each stage. Thus

$$w = \text{work done} = \int_{V_1}^{V_2} P\,dV$$

$$= 2.303 \text{ n RT} \log \frac{V_2}{V_1}$$

$$= 2.303 \text{ nRT} \log \frac{p_1}{P_2}$$

given date :

$$P_1 = 2 \text{ atm}$$
$$P_2 = 2 \text{ atm}$$
$$T = 300 \text{ K}$$
$$n = \frac{16}{32} = 0.5 \text{mole.}$$

On substituting these values in the above equation, we get

$$w = 2.303 \times 0.5 \times 8.314 \times 300 \log \frac{2}{200}$$

$$= -5744 \text{ J.}$$

Problem 52:

Four moles of an ideal gas are held by a piston under 5 atm pressure and at 273.15 K. The pressure is suddenly released to 0.2 atm and the gas is allowed to expand isothermally. Calculate w, q, ΔE and ΔH for the process.

Solution:

As the pressure has been released suddenly, the expansion must be irreversible.

Here n = 4 moles; P_1 = 5 atm; $T_1 = T_2$ = 273.15 K.

(i) Work done irreversibly

$$w = nRT\left\{1 - \frac{P_2}{P_1}\right\}$$

$$= 4 \times 8.314 \times 273.15 \left\{1 - \frac{0.2}{5}\right\}$$

$$= 4 \times 8.314 \times 273.15 \times 0.96$$

$$= 8720 \text{ J.}$$

(ii) As the ideal gas expands isothermally

$$dT = 0 \text{ and hence } \Delta E = 0; \Delta H = 0.$$

(iii) First law gives

$$\Delta E = q - w = 0$$

$$\therefore q = w\ 8720 \text{ J.}$$

Problem 53:

At low temperature, a nonideal gas follows the relation PV = RT -- a/V where a = 0.3636 Nm⁴ mole ². Calculate the work done by one mole of this gas in expanding form 0.224 litre to 22.4 litre at 400 K. Compare this result with the work done in the corresponding expansion of an ideal gas.

Solution:

For an ideal gas, reversible work done

$$= 2.303 \text{ n RT} \log \frac{V_2}{V_1}$$

V_1 = 0.224 litres mole^{-1} = 22.4 × 10^{-5} m^3 mole^{-1}

P_2 = 22.4 litres mole^{-1} = 22.4 × 10^{-3} m^3 mole^{-1}

R = 8.314 J K^{-1} mole^{-1}

$$W_{rew} = (2.303)\ (1\ \text{mole})\ (8.314\ \text{J K}^{-1}\ \text{mole}^{-1})\ (400\ \text{K}) \times \log \frac{22.4 \times 10^{-3}}{22.4 \times 10^{-5}}$$

$$= (2.303)\ (1\ \text{mole})\ (8.314\ \text{J K}^{-1}\ \text{mole}^{-1})\ (400\ \text{K}) \log 100$$

$$= 15320\ \text{J mole}^{-1}$$

$$= 15.32\ \text{kJ mole}^{-1}.$$

For the non-ideal gas PV = RT – a/V

$$\text{Therefore, the work done} = \int_{V_1}^{V_2} P\,dV = \int_{V_1}^{V_2} \left\{\frac{RT}{V} - \frac{a}{V_2}\right\} dV$$

$$= 2.303\ RT \log \frac{V_2}{V_1} + a\left\{\frac{1}{V_2} - \frac{1}{V_1}\right\}$$

The value of first part is from derived eq. (1) as 15.32 kJ mole^{-1} and of another part can be derived as

$$0.3636\ \text{Nm}^4\ \text{mole}^{-2} \left\{\frac{1}{22.4 \times 10^{-3}\ \text{m}^3\ \text{mole}^{-1}} - \frac{1}{22.4 \times 10^{-5}\ \text{m}^3\ \text{mole}^{-1}}\right\}$$

$$= 1.604\ \text{kJ mole}^{-1}.$$

Thus work done = 15.320 k Joule^{-1} – 1.604 k Joule^{-1}

= 13.716 kJ mole^{-1}.

Problem 54:

What will be the work done when 65.38 g of zinc dissolves in hydrochloric acid in case of an open beaker and closed beaker at 300K.

Solution:

The process of dissolution of zinc in hydrochloric acid can be given as

$$Zn(s) + 2HCI\ (aq) = ZnCl_2(aq) + H_2(g)$$

From this equation it is clear that for each gram atom of zinc dissolved we will obtain 1 mole of Hydrogen gas. As a result of liberated gas, on the surrounding atmosphere the work performed will be equal to PΔV. If the initial volume of the system is neglected in comparison to the total volume of the gas produced and the gas shows ideal behaviour, then the work done in the open beaker is

$$W = P\Delta V \approx PV_{H2}$$

$$= nH2\ RT \qquad [\because T = 300\ k\ R = 8.314\ Jk^{-1}\ mole^{-1}]$$

$$= (1\ mole)\ (8.314\ JK^{-1}\ mol^{-1})\ (300\ K)$$

$$= 2.4942\ kJ.$$

If the reaction takes place in a closed beaker there is change in the volume *i.e.*, ΔV = 0 hence work done is zero.

Problem 55:

Calculate the work done when one mole of sulphur dioxide gas expands isothermally and reversibly at 300 k from 2.46 × 10^{-3} m^3 to 24.6 × 10^{-3} m^3. Assuming that the gas obeys

(a) ideal gas equation and

(b) van der Waals equation

$$\left\{\begin{matrix} a = 0.6799\ nm^4\ mole^{-2} \\ b = 0.564 \times 10^{-3}\ mole^{-1} \end{matrix}\right\}$$

Solution:

(a) The reversible work done by 1 mole of gas when it expands form

$$2.64 \times 10^{-3}m^3 \text{ to } 24.6 \times 10^{-3}m^3 = 2.303\ RT \log \frac{V_2}{V_1}$$

$$= (2.303)\ (1\ mole)\ (8.314\ J\ K^{-1}\ mol^{-1})\ (300\ K) \times \log 10$$

$$\left\{\because \log \frac{V_2}{V_1} = \log 10\right\}$$

$$= (2.303)\ \ (8.314)\ (300\ K)\ \ \log 10\ J\ mol^{-1}$$

$$= 5744\ J\ mol^{-1}$$

(b) When the gas obeys van der Waal's equation, work done by 1 mole of gas is

$$= 2.303\ RT \log \frac{V_2 - b}{V_1 - b} + a\left\{\frac{1}{V_2} - \frac{1}{V_1}\right\}$$

After substituting the values of various quantities and solving, the work done is

$$= 5794 \text{ J mol}^{-1} - 252.4 \text{ J mol}^{-1}$$

$$= 5541.6 \text{ J mol}^{-1}$$

$$= 5.542 \text{ kJ mol}^{-1}.$$

Problem 56:

One mole of a van der Walls's gas is allowed to expand isothermally and reversibly from a volume of 1 litre to 50 litre at 0°C. Calculate w, q, Δ E and ΔH. van der Waal's constants are a = 6.5 atm l²mole⁻² = 0.056 litre mole⁻¹ and R = 0.082 l atom deg⁻¹ mole⁻¹.

Solution:

Here $V_1 = 1$ litre,

$V_2 = 50$ litres, n = 1 mole

$T = 273 + 0 = 273°k$,

$R = 0.082$ l-atm deg^{-1} $mole^{-1}$

$a = 6.5$ atm l^2 $mole^{-2}$,

$b = 0.056$ litre $mole^{-1}$

$$w = 2.303 \text{ nRT} \log \left(\frac{V_2 - nb}{V_1 - nb}\right) + an^2 \left(\frac{1}{V_2} - \frac{1}{V_1}\right)$$

$$= 2.303 \times 1 \times 0.082 + 273 \left[\log(50 - 0.056) - \log(1 - 0.056)\right.$$

$$+ 6.5\left(\frac{1}{50} - \frac{1}{1}\right)$$

$$= 82.47 \text{ l atm.}$$

$$\Delta E = -an^2 \left(\frac{1}{V_2} - \frac{1}{V_1}\right) = 6.5 \times \frac{49}{50}$$

$$q = 2.303\text{nRT} \log \frac{V_2 - b}{V_1 - b}$$

$$= \Delta E + w = 884 \text{ l atm.}$$

Problem 57:

The van der Waal's constant a and b for hydrogen in litre atmosphere units are 0.246 and 2.67 × 10⁻² respectively. Calculate the inversion temperature of hydrogen.

Solution:

we know $T_i = 2a/Rb$

or $T_i = \dfrac{2 \times 0.246}{0.0821 \times 0.0267}$

$= 224.5°k = (224.5 - 273) = - 48.5°C.$

Problem 58:

For equation $(P \_ a/V^2)\ V = RT$, prove that (i) dP is an exact differential, (ii) P is a state function and

$$\text{(iii)} \left(\frac{\partial P}{\partial T}\right)_V \left(\frac{\partial T}{\partial V}\right)_P \left(\frac{\partial V}{\partial T}\right)_T + 1 = 0.$$

Solution:

(i) For dP to be an exact differential, it is be proved that

$$\frac{\partial^2 P}{\partial V \partial T} = \frac{\partial^2 P}{\partial T \partial V} \quad ...(1)$$

$$(P + a/V^2)\ V = RT$$

or $$P = \frac{RT}{V} - \frac{a}{V^2} \quad ...(2)$$

When the above equation is differentiated with respect to V at constant temperature T, we obtain

$$\left(\frac{\partial P}{\partial V}\right)_T = -\frac{RT}{V^2} + \frac{2a}{V^3}$$

When the above equation is differentiated with respect to T at constant V, we obtain

$$\frac{\partial^2 P}{\partial T \partial V} = -\frac{RT}{V^2} \quad ...(3)$$

When equation (2) is differentiated with respect to T at constant volume V, we obtain

$$\left(\frac{\partial P}{\partial T}\right)_V = \frac{R}{V}$$

When the above equation is further differentiated with respect to V at constant T, we obtain

$$\frac{\partial^2 P}{\partial T \partial V} = -\frac{R}{V^2} \quad ...(4)$$

From equations (3) and (4), we get

$$\frac{\partial^2 P}{\partial T\, \partial V} = \frac{\partial^2 P}{\partial T\, \partial V} = -\frac{R}{V^2} \qquad ...(5)$$

Thus, we have proved condition, (1). This shows that dP is an exact differential.

(ii) In the said question, first of all a change occurs in volume at constant temperature followed by a change in temperature at constant volume, thereby giving the final equation (3).

Again, a change in temperature is considered at constant volume, followed by a change in volume at constant temperature, thereby giving the final equation (4). Both equations (3) and (4) are same, thereby revealing that the final change in P remains the same irrespective of the difference in intermediate changes. Therefore, P is a state function as changes in P have been found to depend upon the intial and final states of the system.

(iii) Equation (2) is as follows :

$$P = \frac{RT}{V} - \frac{a}{V^2}$$

When the above equation is differentiated, we obtain :

$$dP = \frac{R}{V}\, dT - \frac{RT}{V^2}\, dV + \frac{2a}{V^3}\, dV$$

$$= \frac{R}{V}\, dT - \left(\frac{RT}{V^2} - \frac{2a}{V^3}\right) dV \qquad ...(6)$$

If volume is constant, dV = 0 : equation (6) becomes as follows :

$$dP = \frac{R}{V}\, dT \text{ or } \left(\frac{\partial P}{\partial V}\right)_V = \frac{R}{V} \qquad ...(7)$$

If pressure is constant, dP = 0; equation (6) becomes as follow :

$$\frac{R}{V}\, dT - \left(\frac{RT}{V^2} - \frac{2a}{V^3}\right) dV = 0$$

or

$$\left(\frac{\partial T}{\partial V}\right)_P = \left(\frac{RT}{V^2} - \frac{2a}{V^3}\right) \bigg/ \frac{R}{V} \qquad ...(8)$$

If temperature is constant, dT = 0; equation (6) becomes as follows:

$$\left(\frac{\partial T}{\partial V}\right)_T = -\frac{1}{\left(\frac{RT}{V^2} - \frac{2a}{V^3}\right)} \qquad ...(9)$$

On multiplying equations (7), (8) and (9), we obtain

$$\left(\frac{\partial p}{\partial T}\right)_V \left(\frac{\partial T}{\partial V}\right)_P \left(\frac{\partial V}{\partial P}\right)_T = -\frac{R}{V} \times \frac{\left(\frac{RT}{V^2} - \frac{2a}{V^3}\right)}{\left(\frac{RT}{V^2} - \frac{2a}{V^3}\right)} \times \frac{1}{R} = -1$$

or $$\left(\frac{\partial p}{\partial T}\right)_V \left(\frac{\partial T}{\partial V}\right)_P \left(\frac{\partial V}{\partial P}\right)_T + 1 = 0$$

This is cyclic rule.

Problem 59:

For an ideal gas PV = RT, prove that dP is an exact differential.

Solution:

We know PV = RT

or $$P = \frac{RT}{V} \text{ (R is a constant)} \quad ...(1)$$

On differentiating with respect to T, taking V constant, we obtain

$$\left(\frac{\partial P}{\partial T}\right)_P = \frac{RT}{V} \quad ...(2)$$

Again, on differentiating with respect to V, we obtain

$$\frac{\partial^2 P}{\partial V \partial T} = -\frac{R}{V^2} \quad ..(3)$$

On differentiating (1) with respect to volume, taking T constant, we obtain

$$\left(\frac{\partial P}{\partial V}\right)_T = -\frac{RT}{V^2} \quad ...(4)$$

Again, on differentiating above equation with respect to temperature taking V as constant, we obtain

$$\frac{\partial^2 P}{\partial T \partial V} = -\frac{R}{V^2} \quad ...(5)$$

From equations (3) and (5), we obtain

$$\frac{\partial^2 P}{\partial T \partial V} = \frac{\partial^2 P}{\partial T \partial V}$$

Thus dP is an exact differential.

Problem 60:

For an ideal gas PV = RT, prove that DT is an exact differential.

Solution:

$$PV = RT \qquad ...(1)$$

In order to prove that dT is a perfect differential, we have to prove that

$$\frac{\partial^2 T}{\partial V \partial P} = \frac{\partial^2 T}{\partial P \partial V}$$

On differentiating Eq. (1), we obtain

$$\left(\frac{\partial T}{\partial V}\right)_P = \frac{P}{R}; \text{i.e.,} \ \frac{\partial^2 T}{\partial P \partial V} = \frac{1}{R}$$

Again

$$\left(\frac{\partial T}{\partial P}\right)_V = \frac{V}{R}; \text{i.e.,} \ \frac{\partial^2 T}{\partial V \partial P} = \frac{1}{R}$$

Therefore, $\dfrac{\partial^2 T}{\partial V \partial P} = \dfrac{\partial^2 T}{\partial P \partial V}$

Thus, dT is an exact differential.

Problem 61:

For an ideal gas (PV = nRT) show that (1/T) is an integrating factor for dw = P dV.

Solution:

$$dw = P\, dV \qquad ...(1)$$

$\because$ $$PV = nRT$$

$\therefore$ $$p\, dV + V\, dP = nR\, dT$$

and $$dV = \frac{nR}{P} dT - \frac{V}{P} dP \qquad ...(2)$$

On substituting the expression for dV in equation (1), we obtain

$$dw = nR\, dT - V\, dP = nR\, dT - \frac{nRT}{P} dP$$

Now $$\frac{\partial}{\partial P}(nR)_T = 0$$

and $$\frac{\partial}{\partial T}\left(-\frac{nRT}{P}\right)_P = -\frac{nR}{P}$$

Hence, dw is inexact differential.

Suppose f = 1/T be an integrating factor. Then

$$f \,.\, dw = \frac{1}{T} nR\, dT - \frac{nR}{P} dP$$

$$\therefore \quad \frac{\partial}{\partial P}\left[\frac{nR}{T}\right]_T = 0$$

and $\frac{\partial}{\partial T}\left[\frac{nR}{P}\right]_T = 0$

Thus f. dw$\frac{nR}{T}$dT – $\frac{nR}{P}$dP is an exact differential while 1/T is an integrating factor.

Problem 62:

Calculate the maximum work obtainable by (i) the isothermal expansion and (ii) the adiabatic expansion of 2 mol of on ideal gas initially at 25° C form 10 L to 20 L. Assume Cv = (5/2) R.

Solution:

(i) Maximum work is obtained in reversible expansion. Hence

$$w = -\, nRT \ln \frac{V_2}{V_1}$$

$$= -\,(2 \text{ mol})\,(8.314 \text{ J K}^{-1} \text{ mol}^{-1})\,(298 \text{ K})\; 2.303 \log \frac{20L}{10L}$$

$$= -\, 3435 \text{ J}$$

(ii) Firstly, we calculate the final temperature using the expression

$$T_2\, V_2^{\,\gamma-1} = T_1\, V_1^{\,\gamma-1}$$

where $\gamma = \frac{C_P}{C_V} = \frac{(7/2)R}{(5/2)R} = \frac{7}{5}$

Hence, $T_2 = (298K)\left(\frac{10L}{20L}\right)^{(7/5)-1}$

$$= (298 \text{ K})\,(0.7578) = 225.8 \text{ K}$$

Now $W_{adi} = \Delta E$

$$= n\, C_{V},\, m(T_2 - T_1)$$

$$= (2\text{mol})(2.5 \times 8.314 \text{ JK}^{-1}\text{mol}^{-1})(225.8\text{K} - 298\text{K})$$

$$= -3001 \text{ J.}$$

Problem 63:

How much heat is required to raise the temperature of 1 mole oxygen from 300K to 1300 K at constant pressure.

$$C_p = 6.095 + 3.325 \times 10^{-3} T - 1.017 \times 10^{-6} T^2.$$

Solution:

$$H_2 - H_1 \int_{T_1}^{T_2} C_p \, dT$$

$$T_1 = 300 \text{ k}, T_2 = 100 \text{ k}$$

$$\therefore \quad H_{1300} - H_{300}$$

$$= \int_{300}^{1300} \left(6.095 + 3.253 + 10^{-3} T - 1.017 \times 10^{-6} T^2\right)$$

[Substituting the value of C_p]

$$\because \quad H_{1300} - H_{300} = 6.095 (1300 - 300) + \frac{3.253 \times 10^{-3}}{2}$$

$$\times [(1300)^2 - (300)^2] - 1/3 \times 1.017 \times 10^{-6} [(1000)^2 - (300)^2]$$

$$= 6095 + 2602.4 - 735.63$$

$$= 7961.77 \text{ cal/mole.}$$

Problem 64:

Show that

$$\left(\frac{\partial \alpha}{\partial P}\right)_T + \left(\frac{\partial \beta}{\partial T}\right)_P = 0$$

where α and β are coefficient of thermal expansion and coefficient of compressibility, respectively.

Solution:

Since V is a state function, we can write

$$\frac{\partial^2 V}{\partial T \, \partial P} = \frac{\partial^2 V}{\partial P \, \partial T}$$

or
$$\frac{\partial}{\partial T}(-V\beta)_P = \frac{\partial}{\partial T}(V\alpha)_T$$

Carrying out the differentiation, we get

$$-\beta\left(\frac{\partial V}{\partial T}\right)_P - V\left(\frac{\partial \beta}{\partial T}\right)_P = \alpha\left(\frac{\partial V}{\partial P}\right)_T + V\left(\frac{\partial \alpha}{\partial P}\right)_T$$

that is $$\left(\frac{\partial \alpha}{\partial P}\right)_T + \left(\frac{\partial \beta}{\partial T}\right)_P = -\frac{\beta}{V}\left(\frac{\partial V}{\partial T}\right)_P - \frac{\alpha}{V}\left(\frac{\partial V}{\partial P}\right)_T$$

$$= -\frac{\beta}{V}(\alpha V) - \frac{\alpha}{V}(-\beta V) = 0$$

Problem 65:

Show that Joule Thomson coefficient, $\mu_{JT} = 0$ for an ideal gas. Comment on the liquefaction of an ideal gas.

Solution:

By definition, Joule-Thomson coefficient is defined as

$$\mu_{JT} = \left(\frac{\partial T}{\partial P}\right)_H$$

Using the cyclic rule

$$\left(\frac{\partial T}{\partial P}\right)_H \left(\frac{\partial P}{\partial H}\right)_T \left(\frac{\partial H}{\partial T}\right)_P + 1 = 0$$

we get $$\left(\frac{\partial T}{\partial P}\right)_H = -\frac{1}{\left(\frac{\partial P}{\partial H}\right)_T \left(\frac{\partial H}{\partial T}\right)_P} = -\frac{\left(\frac{\partial H}{\partial P}\right)_T}{C_P}$$

Now using, thermodynamic equation of state

$$\left(\frac{\partial T}{\partial P}\right)_H = V - T\left(\frac{\partial V}{\partial T}\right)_P$$

we get $$\mu_{JT} = \frac{V - T\left(\frac{\partial V}{\partial T}\right)_P}{C_P}$$

Now, for an ideal gas

$$V = \frac{nRT}{P}$$

Hence, $$\left(\frac{\partial V}{\partial T}\right)_P = \frac{nR}{P} = \frac{V}{T}$$

Thus $$\mu_{JT} = \frac{V - T\left(\frac{V}{T}\right)}{C_P} = 0$$

The ideal gas cannot be liquefied as there will not occur any change in temperature when the gas undergoes Joule-Thomson experiment.

Problem 66:

Calculate the work down when 50 g of iron dissolves in hydrochloric acid at 25oC in :

(i) a closed vessel and

(ii) an open beaker.

(iii) In which case would maximum amount of work be done ?

Solution:

$$\text{Amount of iron} = \frac{50\text{ g}}{55.85\text{ g mol}^{-1}} = \frac{50}{55.85}\text{ mol}$$

(i) In a closed vessel, $\Delta V = 0$. Hence,

$$dw = -\,PdV = 0$$

that is, no work is involved when iron dissolves in hydrochloric acid in a closed vessel.

(ii) The reaction between iron and hydrochloric acid is

$$Fe + 2HCl \rightarrow FeCl_2 + H_2(g)$$

Thus, the liberated hydrogen gas pushes the atmospheric air and thus involves the work of expansion. In this case, we will have

$$w = -\,P_{ext}\,\Delta V$$
$$= -\,P_{ext}\,V_{H2}$$

where V_{H2} is the volume of hydrogen released. According to the ideal gas law, we will have

$$w = -\,P_{ext}\,(n_{H2}RT/P_{ext})$$
$$= -\,n_{H2}RT$$
$$= -\,(50/55.85)\text{ mol }(8.314\text{ J K}^{-1}\text{ mol}^{-1})\,(298\text{ K})$$
$$= -\,2.2\text{ kJ}$$

Problem 67:

Since CV = $(\partial q/\partial T)$ V by definition, one often writes without restriction dV = $C_V dT$. This is generally not true. Explain why.

Solution:

If we take U = f)V, T), then we will have

$$dU = \left(\frac{\partial U}{\partial V}\right)_T dV + \left(\frac{\partial U}{\partial T}\right)_V dT$$

$$= \left(\frac{\partial U}{\partial V}\right)_T . dV + C_V\, dT$$

Thus, if one writes $dV = C_V\, dT$ without restriction then this statement is not complete. This is applicable for a system in which $(\partial U/\partial V)_T = 0$. An ideal gas is such an example. Alternatively, the system may be undergoing change at constant volume, for which, $(\partial U/\partial V)_T\, dV$ is zero and thus $dV = C_V\, dT$.

Problem 68:

Calculate the work done by the system when 2 mol of an ideal gas expand from 0.01 m^3 to 0.1 m^3 at 300 k isothermally :

(i) into a vacuum, and

(ii) reversibly.

Explain how the surroundings can be restored only in one case and not in the other.

Solution:

(i) When the work is done into a vacuum, the work done by the system is zero since P_{opp} in this case is zero.

(ii) For the reversible isothermal expansion work involved is

$$w = -\, nRT \ln \frac{V_2}{V_1}$$

$$= (2)\ (8.314 \times 300\ J) \ln \frac{0.1}{0.01}$$

$$= -\, 2 \times 8.314 \times 300 \times 2.303\ J$$

$$= -\, 11488\ J = -\, 11.488\ kJ.$$

This surroundings are not disturbed when the expansion is done against vacuum. In case of reversible isothermal expansion, heat is absorbed by the system from the surroundings. Now if we carry out reversible isothermal compression of the system till the original state, then the same quantity or heat is released to the surroundings and thus the latter is restored back to the original state. However, in the first case where expansion is done against vacuum, there will be bet heat received by the surroundings during compression. Hence, in this case, it is not restored back to its original state.

Problem 69:

Show that $PV = nRt$, then

(i) $\dfrac{\partial^2 P}{\partial n \partial V} = \dfrac{\partial^2 P}{\partial V \partial n} = -\dfrac{RT}{V^2}$

(ii) $\dfrac{\partial^2 P}{\partial n \partial T} = \dfrac{\partial^2 P}{\partial T \partial n} = -\dfrac{R}{V}$

(iii) $\left(\dfrac{\partial P}{\partial T}\right)_V \left(\dfrac{\partial T}{\partial V}\right)_P \left(\dfrac{\partial V}{\partial P}\right)_T = -1$

Solution:

(i) $\because$ $PV = nRT$

$\therefore$ $P = \dfrac{nRT}{V}$

or $\dfrac{\partial P}{\partial V} = \dfrac{\partial}{\partial V}\left(\dfrac{nRT}{V}\right)$ [keeping T as constant]

$$= nRT \frac{\partial}{\partial V}\left(\frac{1}{V}\right) = -\frac{nRT}{V^2}$$

$$\frac{\partial^2 P}{\partial n \partial V} = \frac{\partial}{\partial n}\left(\frac{\partial P}{\partial V}\right)$$

$= \left(-\dfrac{nRT}{V^2}\right)$ [Keeping R, T and V as constant]

$$= -\frac{RT}{V^2} \qquad ...(1)$$

Similarly, $PV = nRT$ or $P = \dfrac{nRT}{V}$

$$\frac{\partial P}{\partial n} = \frac{RT}{V}$$

$\dfrac{\partial^2 P}{\partial V \partial N} = \dfrac{\partial}{\partial V}\left(\dfrac{RT}{V}\right)$ [keeping T and R as constant]

$$= -\frac{RT}{V^2} \qquad ...(2)$$

From equations (1) and (2), we obtain

$$\frac{\partial^2 P}{\partial n \partial V} = \frac{\partial^2 P}{\partial V \partial n} = -\frac{RT}{V^2}$$

(ii) $$P = \frac{nRT}{V}$$

$$\frac{\partial P}{\partial T} = \frac{\partial}{\partial T}\left(\frac{nRT}{V}\right) \quad \text{[Keeping n and R as constant]}$$

$$= \frac{nR}{V}$$

$$\frac{\partial^2 P}{\partial n \partial T} = \frac{\partial}{\partial n}\left(\frac{nR}{V}\right) \quad \text{[Keeping V and R as constant]}$$

$$= \frac{R}{V} \qquad ...(3)$$

Again, $$P = \frac{nRT}{V}$$

$$\frac{\partial P}{\partial n} = \frac{RT}{V} \quad \text{[Keeping RT and V as constant]}$$

$$\frac{\partial^2 P}{\partial T \partial n} = \frac{\partial}{\partial n}\left(\frac{RT}{V}\right) \quad \text{[Keeping RT and V as constant]}$$

or $$\frac{\partial^2 P}{\partial T \partial n} = \frac{R}{V} \qquad ...(4)$$

From equations (3) and (4), we obtain

$$\frac{\partial^2 P}{\partial n \partial T} = \frac{\partial^2 P}{\partial T \partial n} = \frac{R}{V}$$

(iii) Again, $$P = \frac{nRT}{V}$$

$$\left(\frac{\partial P}{\partial T}\right)_V = \frac{nR}{V} \qquad ...(5)$$

Again, $$V = \frac{nRT}{P}$$

$$\left(\frac{\partial V}{\partial P}\right)_T = -\frac{nRT}{P} \qquad ...(6)$$

Again, $$T = \frac{PV}{nR}$$

$$\left(\frac{\partial T}{\partial V}\right)_P = -\frac{P}{nR} \qquad \text{...(7)}$$

On multiplying equations (5), (6), and (7), we obtain

$$\left(\frac{\partial P}{\partial T}\right)_V \left(\frac{\partial T}{\partial V}\right)_P \left(\frac{\partial V}{\partial P}\right)_T = \frac{nR}{V} \times -\frac{nRT}{P^2} \times \frac{P}{nR}$$

$$= -\frac{nRT}{PV} = -1 \left[\because 1 = \frac{nRT}{PV}\right]$$